Olympia, USA

13 April 1949

Books are earthquake proof. Fallen Books is a book project that brings together images of toppled books housed in seismically active libraries. The photographs, often taken by librarians, are laid out as a chronological index of libraries in the aftermath of an earthquake and include earthquake names, locations, dates and the relative intensity of the earthquake as measured by the Modified Mercalli Scale. The accompanying captions are quotes from newspapers and librarian's notes.
The Modified Mercalli Scale is a descriptive and graphic alternative to the Richter Scale, that quantifies how strongly an earthquake affects the Earth's surface, humans, man-made objects and nature, on a scale of I through XII. Topographic maps of earthquake areas use the Mercalli Scale color code to indicate how far away from the epicenter the earthquake was felt and at what degree of intensity.
This book in the hands of the reader will always refer to the copies of Fallen Books introduced into the featured libraries. This book is both an archive and a forecast.

Melissa Dubbin and Aaron S. Davidson

San Benito County Free Library

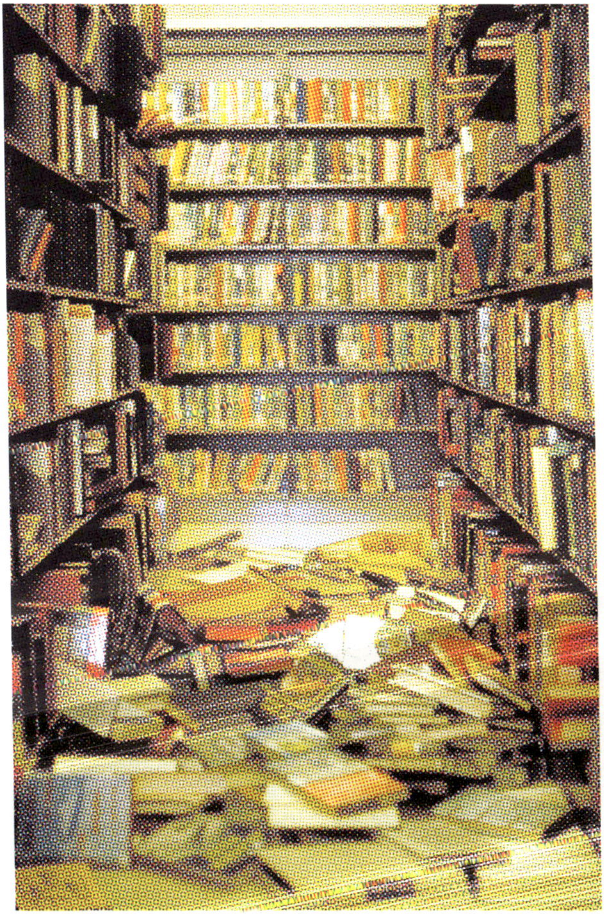

Hollister, USA

Some books fell in the library.

08 April 1961

Anchorage Community College Library

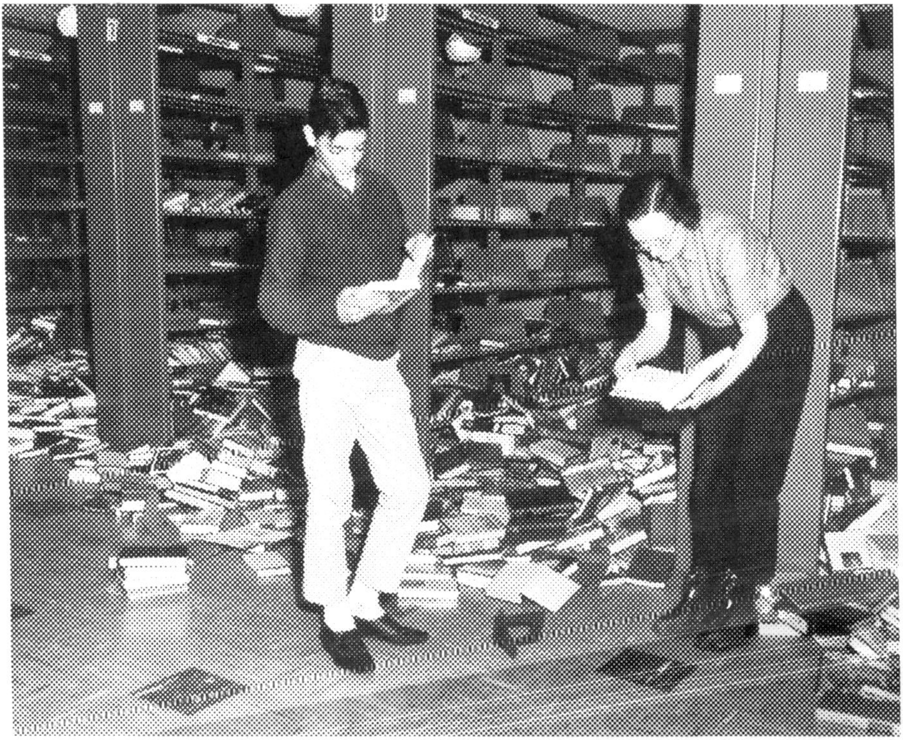

Good Friday/Alaska, USA

27 March 1964

San Fernando Valley State College Library

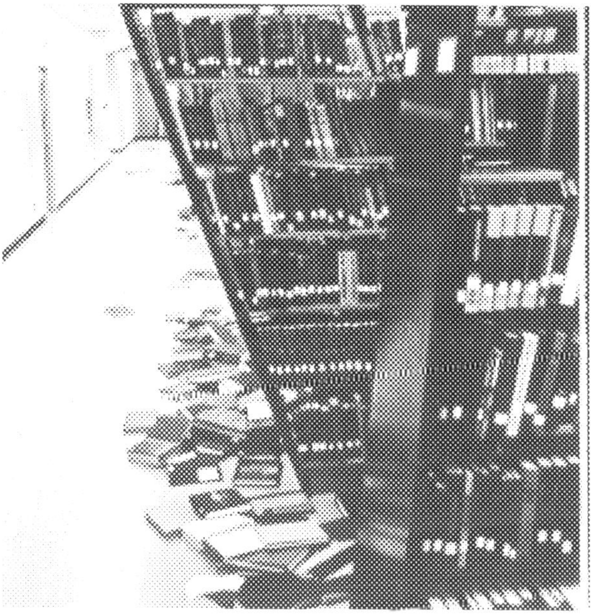

San Fernando/Sylmar, USA

09 February 1971

Los Angeles Public Library

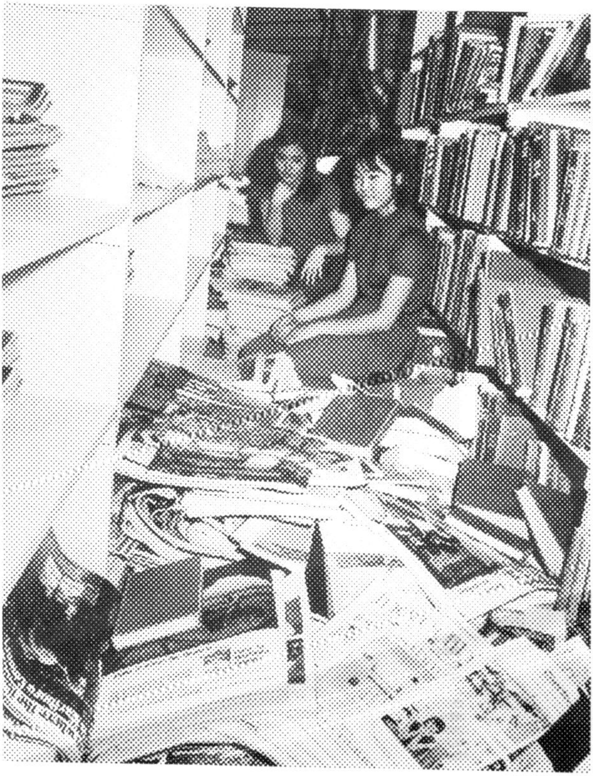

Carol Nishijima and Roberta Ishihara, Gardena High School students, begin the massive job of sorting and re-shelving books at the Los Angeles Public Library's depository at 134 Glendale Boulevard. Several hundred student volunteers have cleaned up more than 100,000 books at the downtown branch and are hard at work on the remaining 61 branches.

09 February 1971

University of California Santa Barbara Library

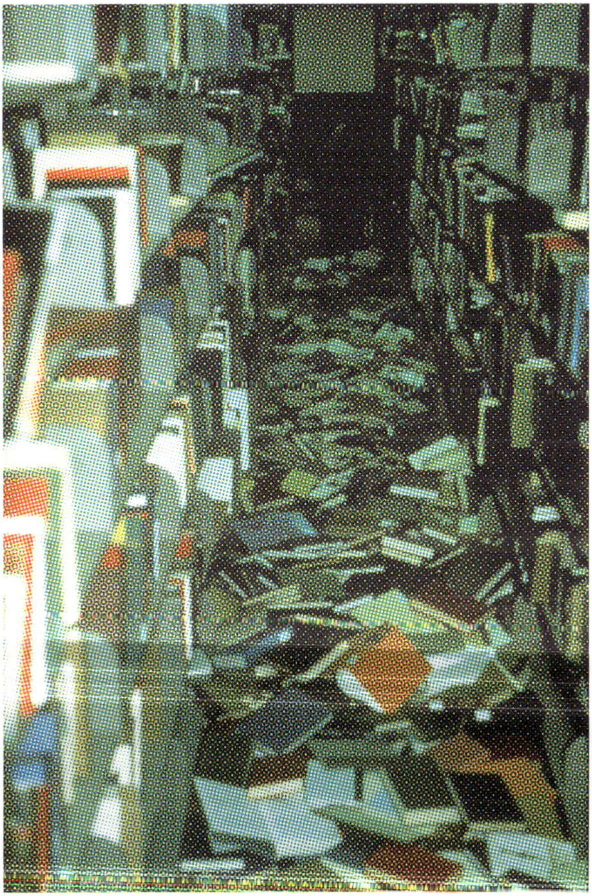

<u>Santa Barbara/Goleta</u>, USA

13 August 1978

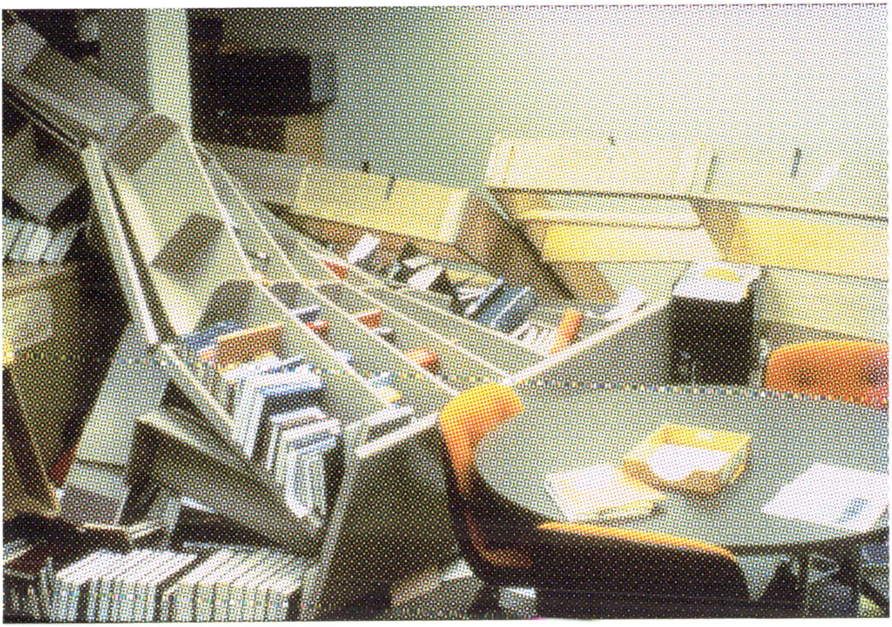

Livermore, USA

24 January 1980

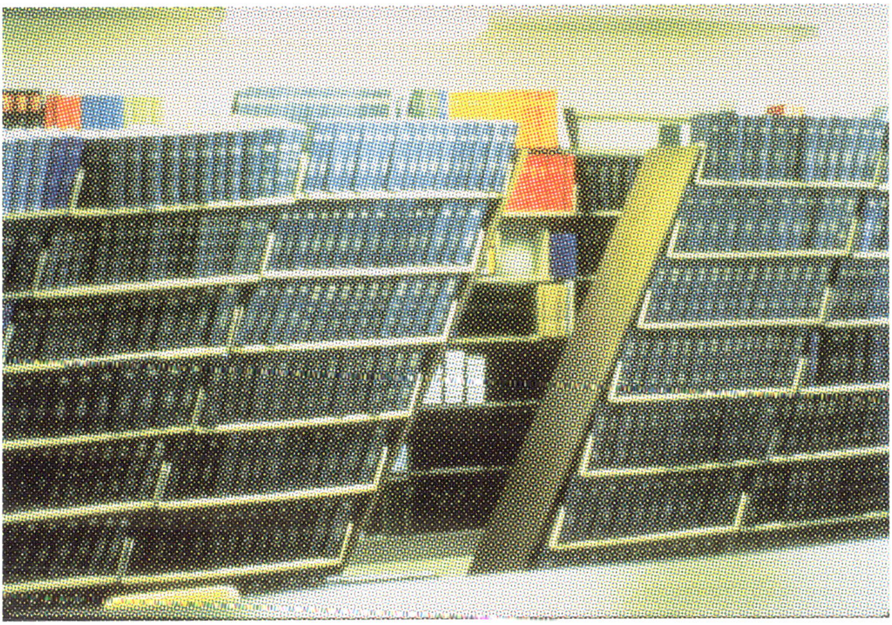

Livermore, USA

24 January 1980

Mammoth Elementary School Library

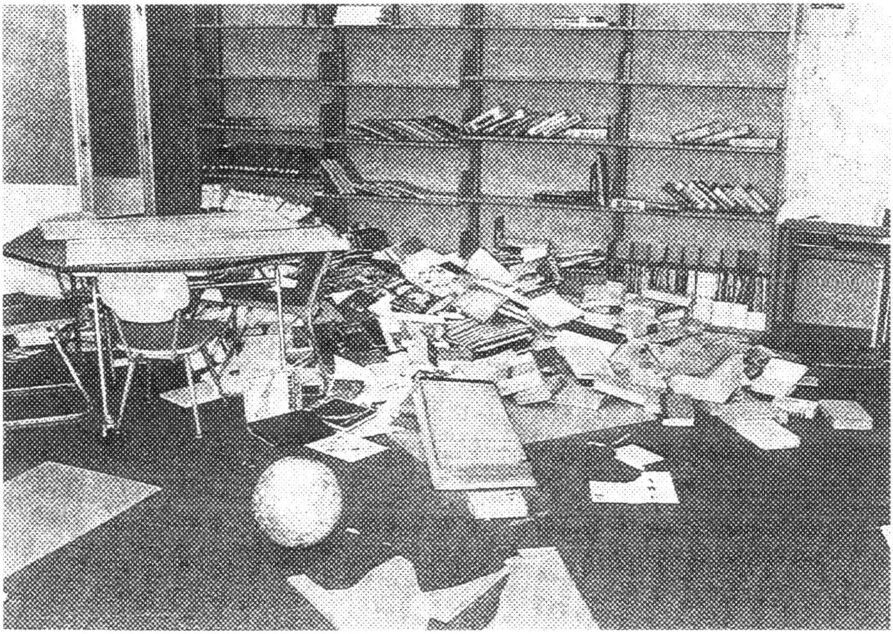

Mammoth Lakes, USA

25 May 1980

Hawaiian Volcano Observatory Library

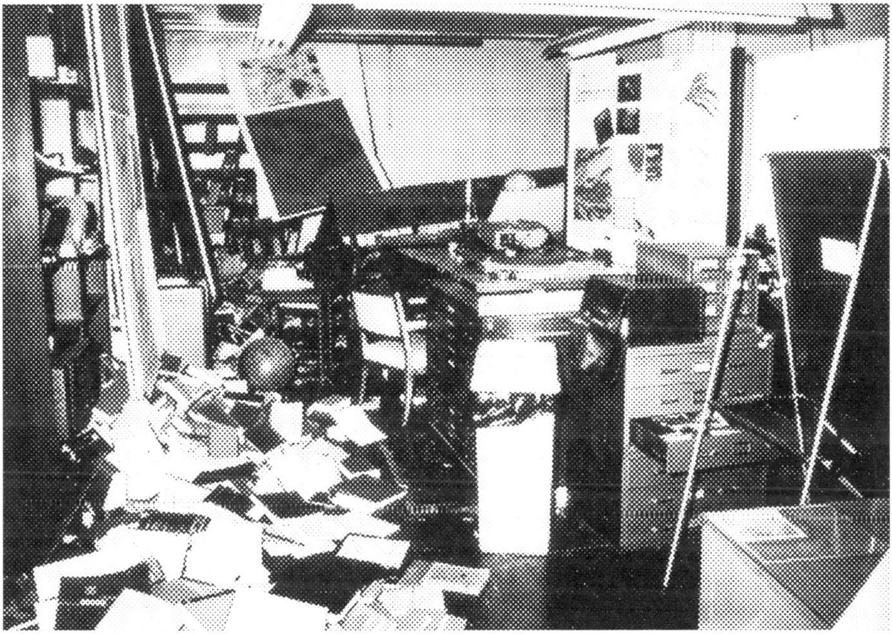

Kaoiki, USA

16 November 1983

University of California Berkeley Library

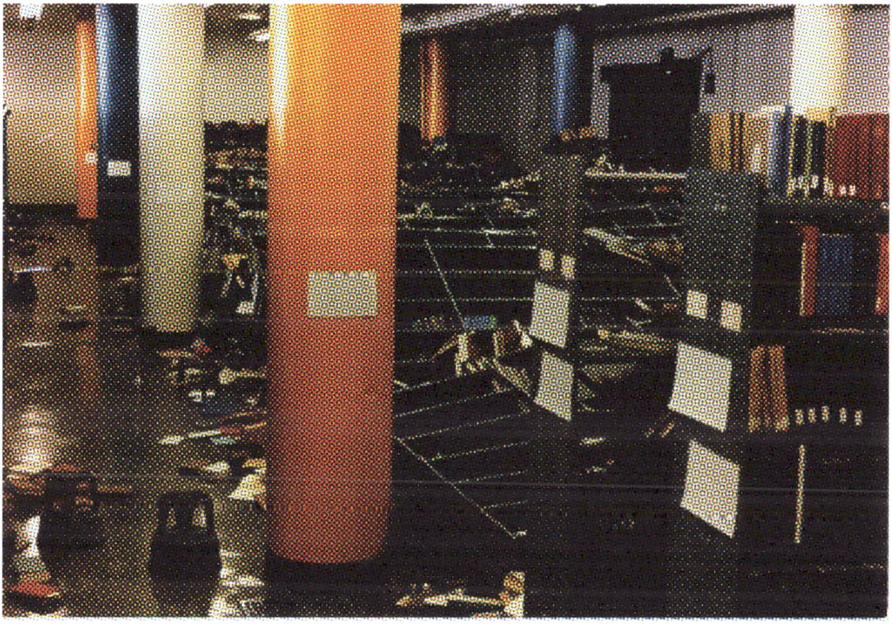

Loma Prieta, USA

17 October 1989

Stanford University School of Business Library

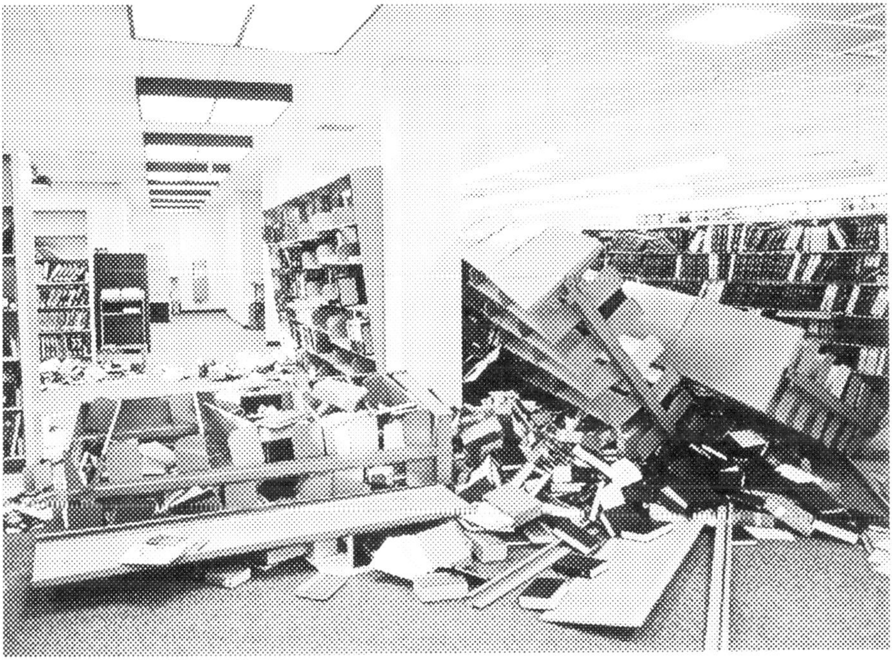

Bookshelves in the Stanford libraries toppled like dominoes. In the days that followed, library volunteers and staff began the daunting task of reshelving some 750,000 volumes.

Claremont Colleges Honnold/Mudd Library

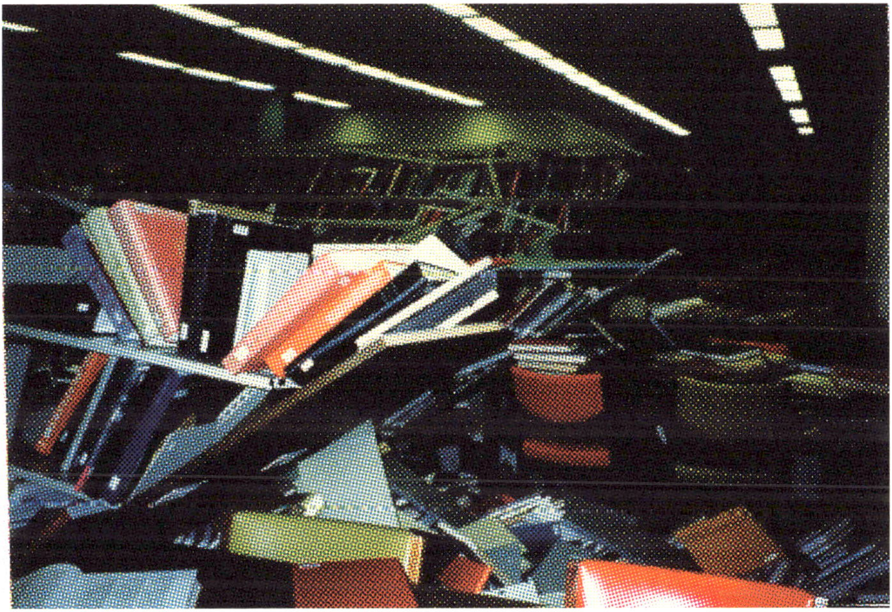

Upland, USA

28 February 1990

Claremont Colleges Honnold/Mudd Library

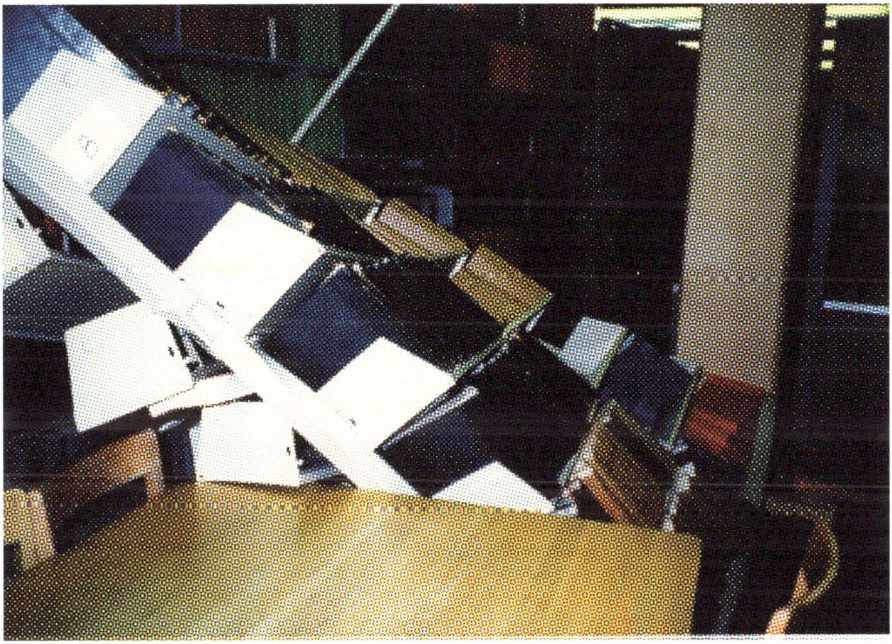

Upland, USA

28 February 1990

Claremont Colleges Honnold/Mudd Library

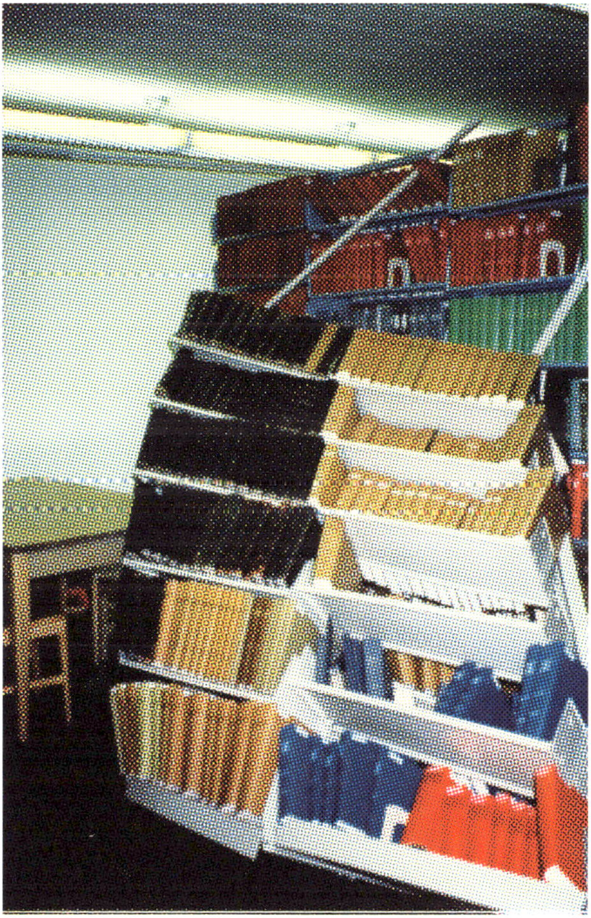

Upland, USA

Note the bracing attached to the tops of stacks did not prevent shelving collapse.

28 February 1990

Claremont Colleges Honnold/Mudd Library

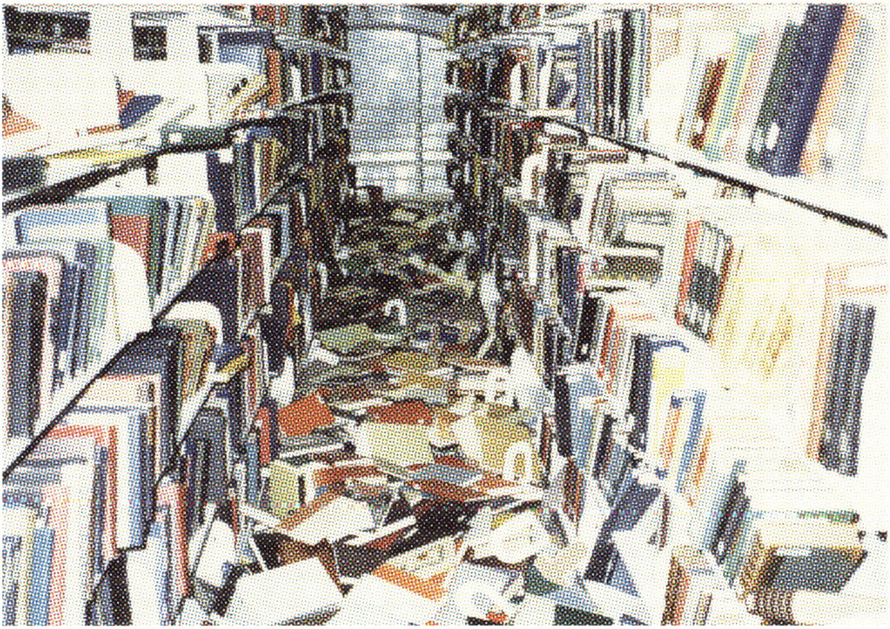

Upland, USA

28 February 1990

Claremont Colleges Honnold/Mudd Library

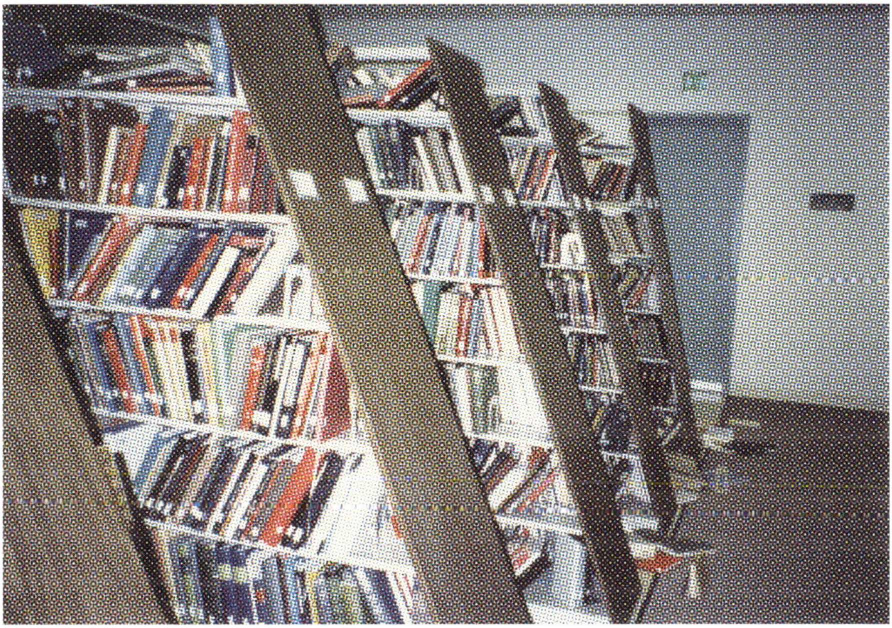

Upland, USA

28 February 1990

Claremont Colleges Honnold/Mudd Library

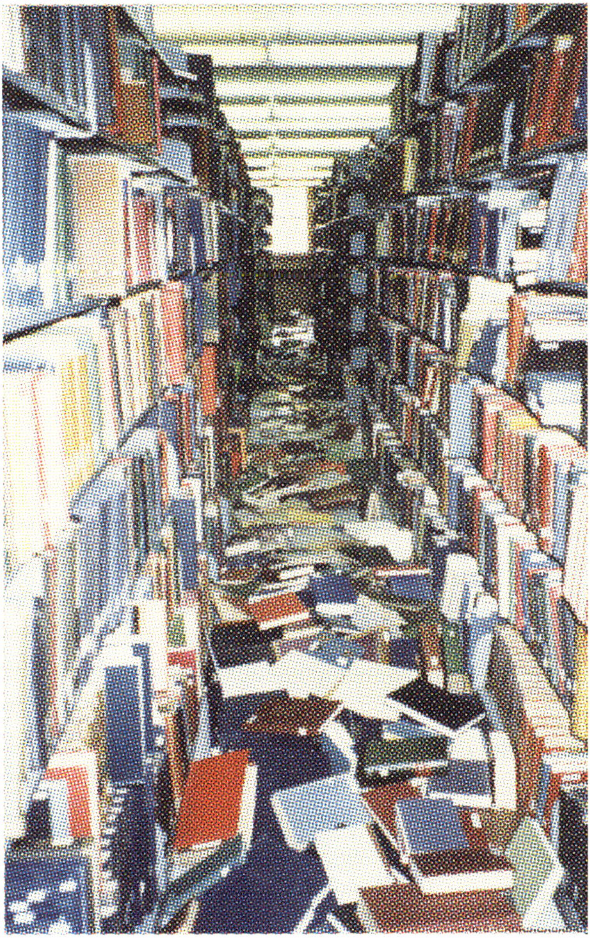

Upland, USA

28 February 1990

Claremont Colleges Honnold/Mudd Library

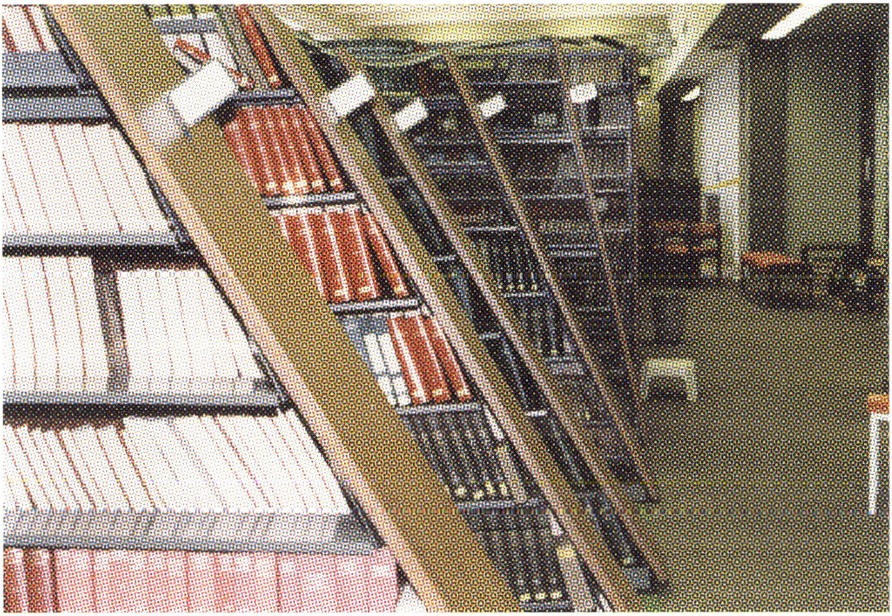

Upland, USA

28 February 1990

Claremont Colleges Honnold/Mudd Library

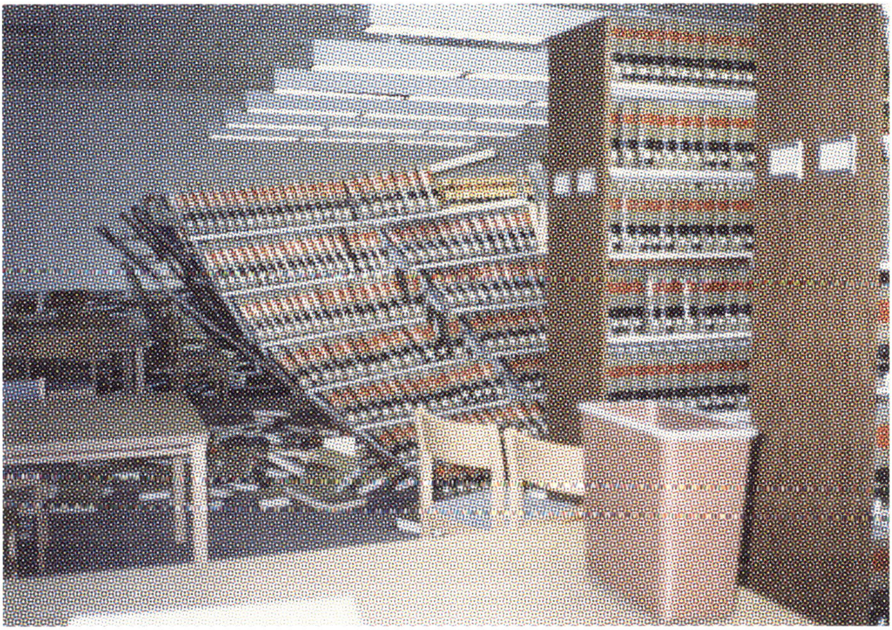

Upland, USA

28 February 1990

University of Redlands Armacost Library

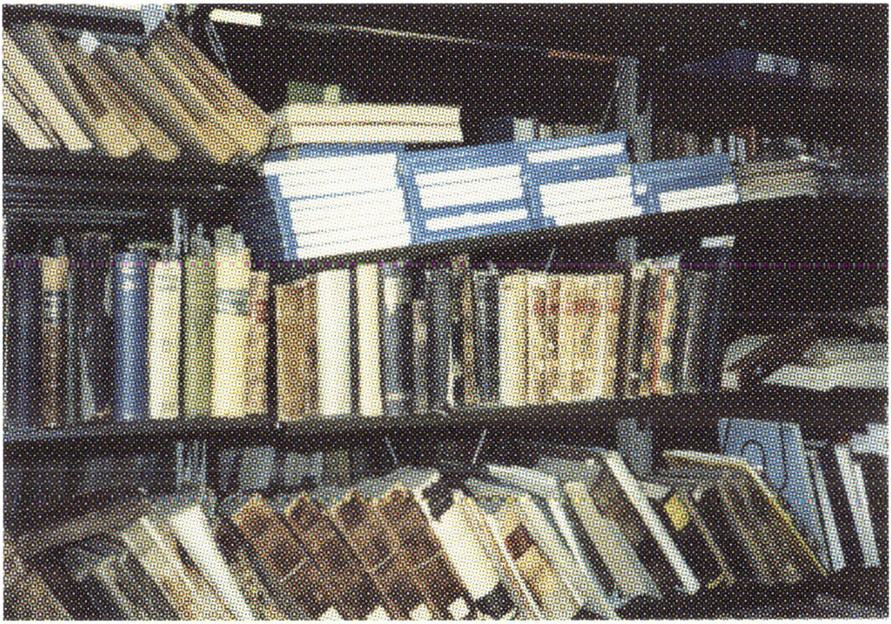

Landers, USA

28 June 1992

University of Redlands Armacost Library

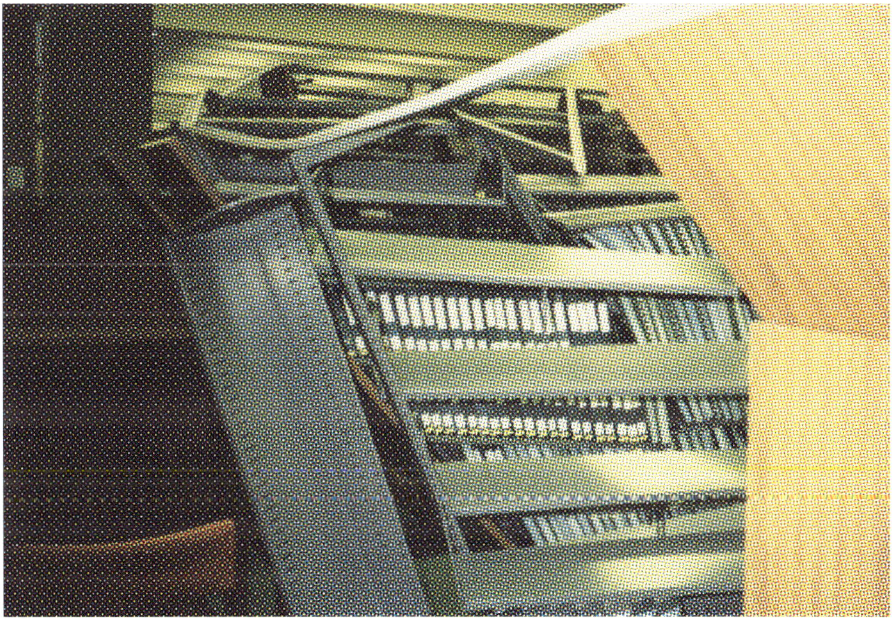

Landers, USA

28 June 1992

University of Redlands Armacost Library

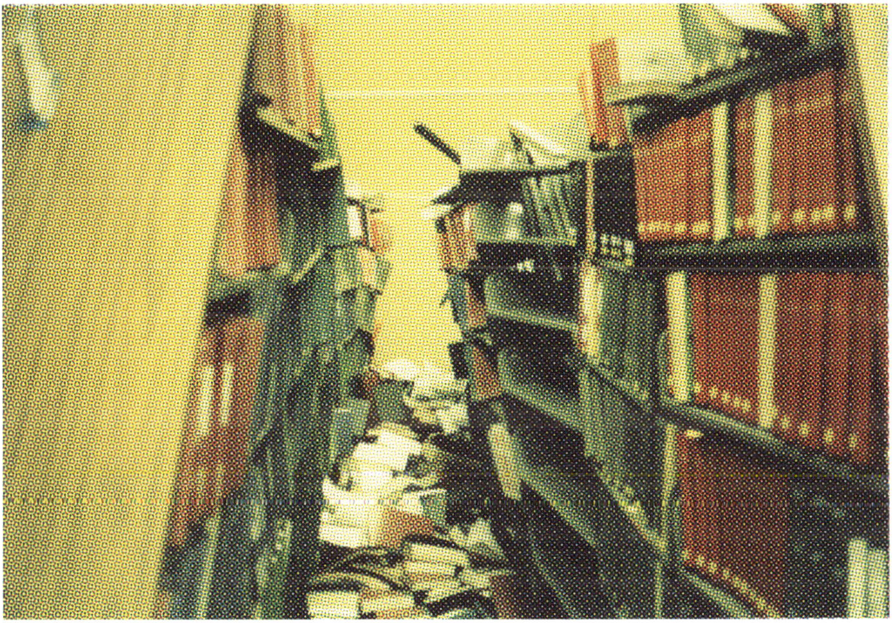

Landers, USA

28 June 1992

University of Redlands Armacost Library

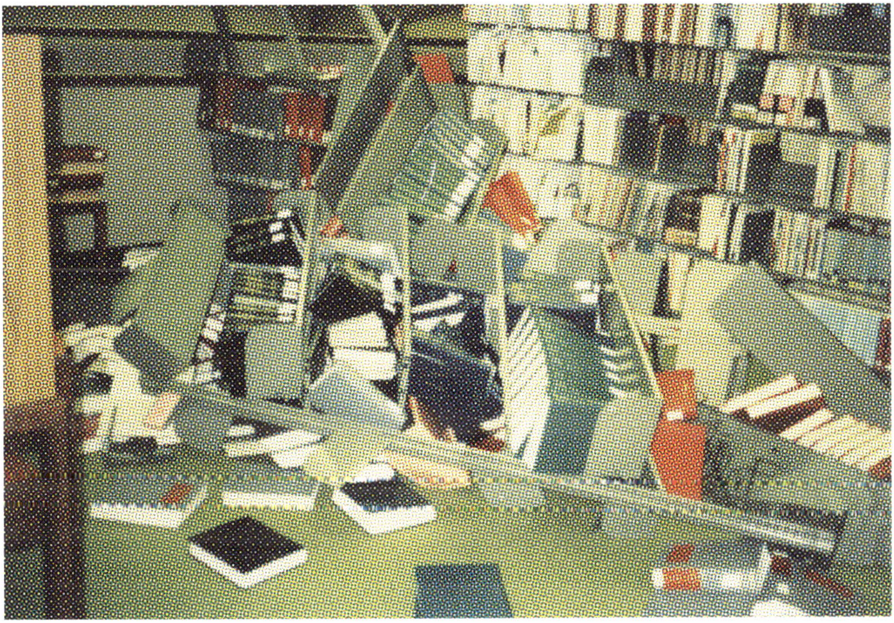

Location is the 4th floor, subjecting the shelving to the most severe shaking in the building.

San Bernardino Valley College Library

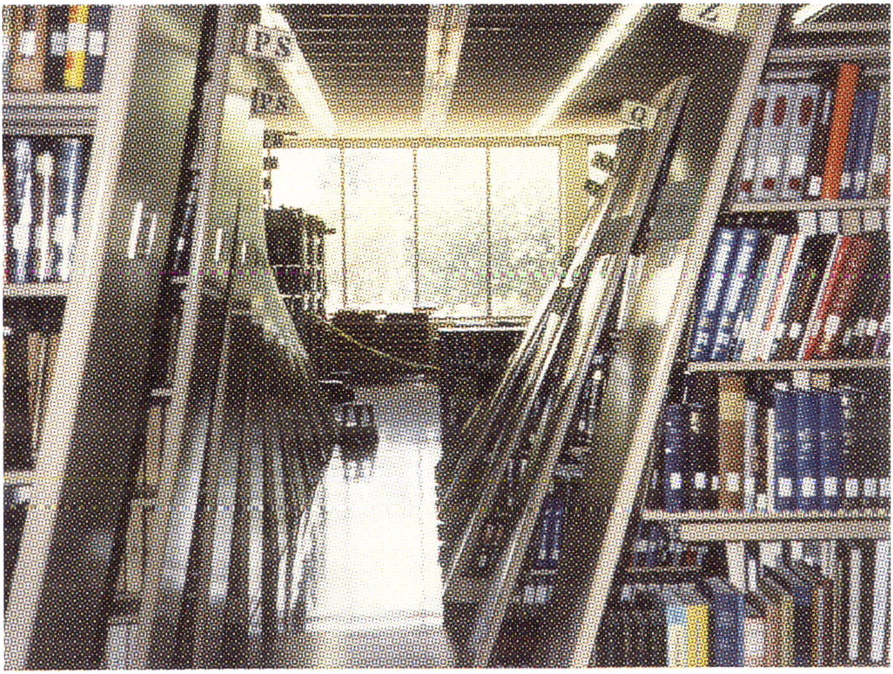

Landers, USA

28 June 1992

San Bernardino Valley College Library

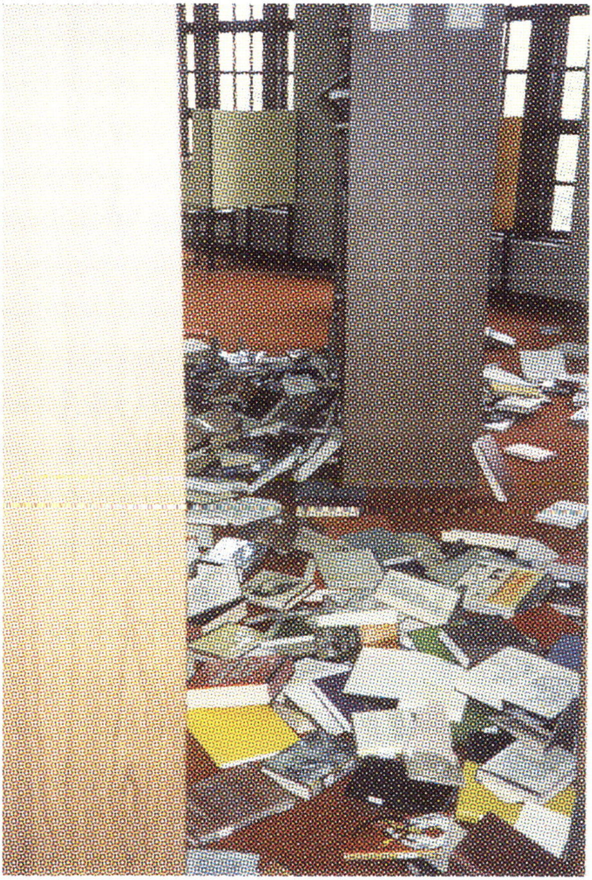

Landers, USA

28 June 1992

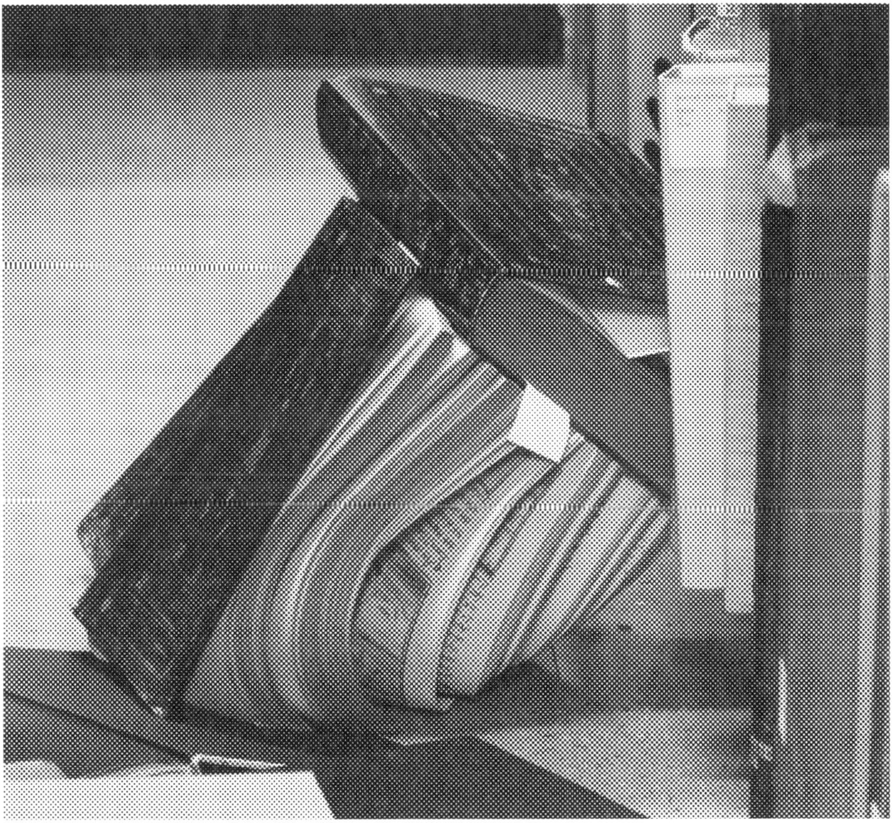

Big Bear, USA

28 June 1992

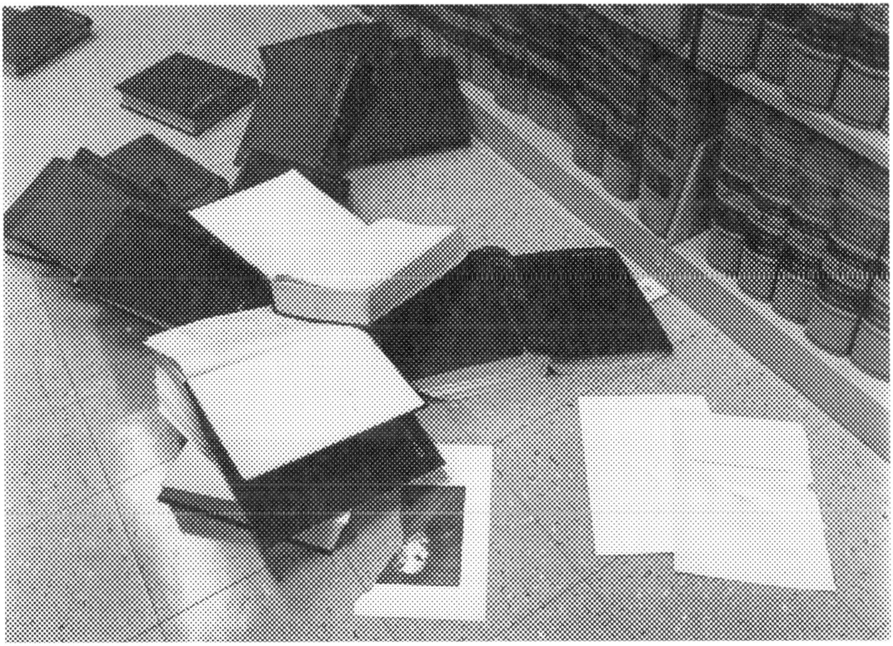

Big Bear, USA

28 June 1992

University of California Riverside Library

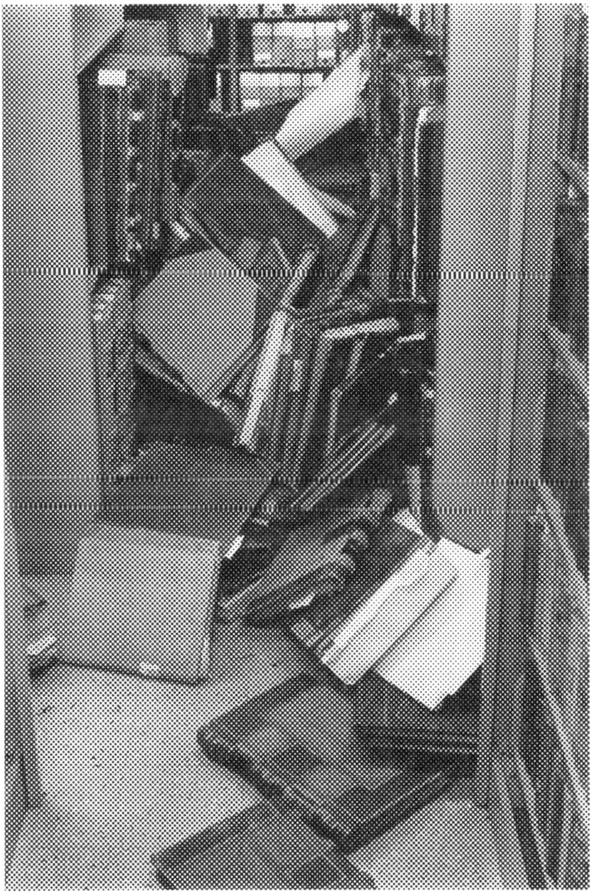

Big Bear, USA

28 June 1992

University of California Riverside Library

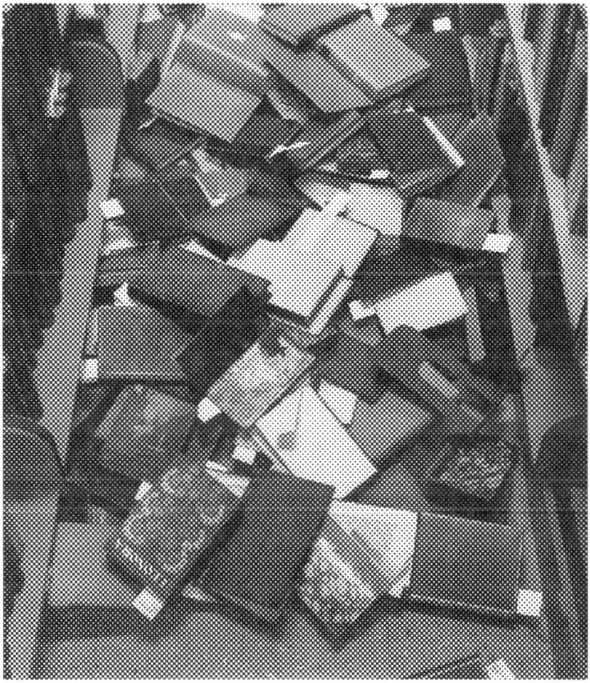

Big Bear, USA

28 June 1992

University of California Riverside Library

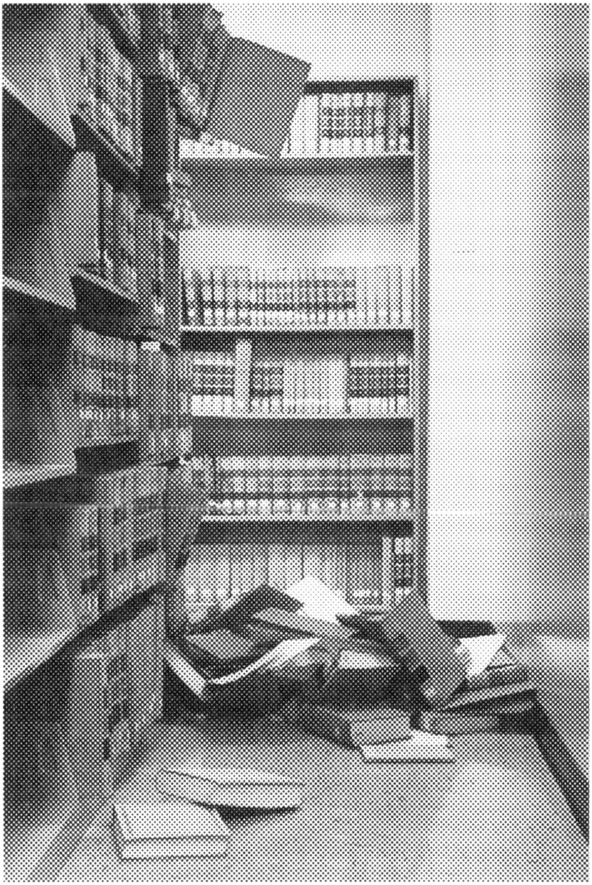

Big Bear, USA

28 June 1992

University of California Riverside Library

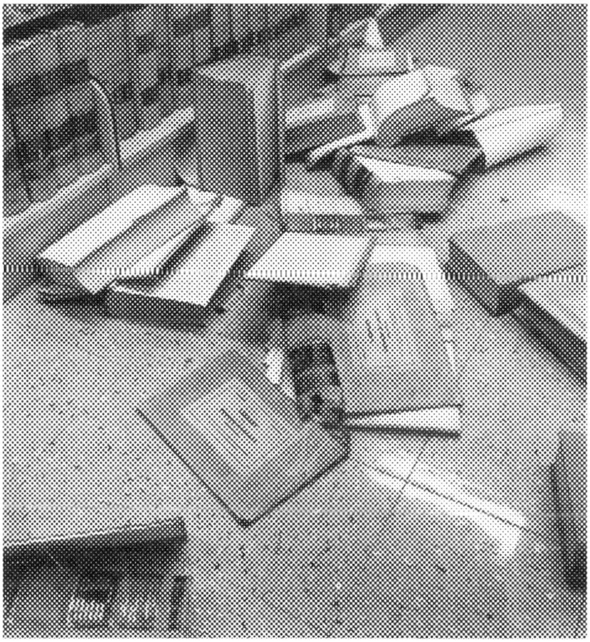

Big Bear, USA

Broken spines and book pages separated from their covers.

28 June 1992

University of California Riverside Library

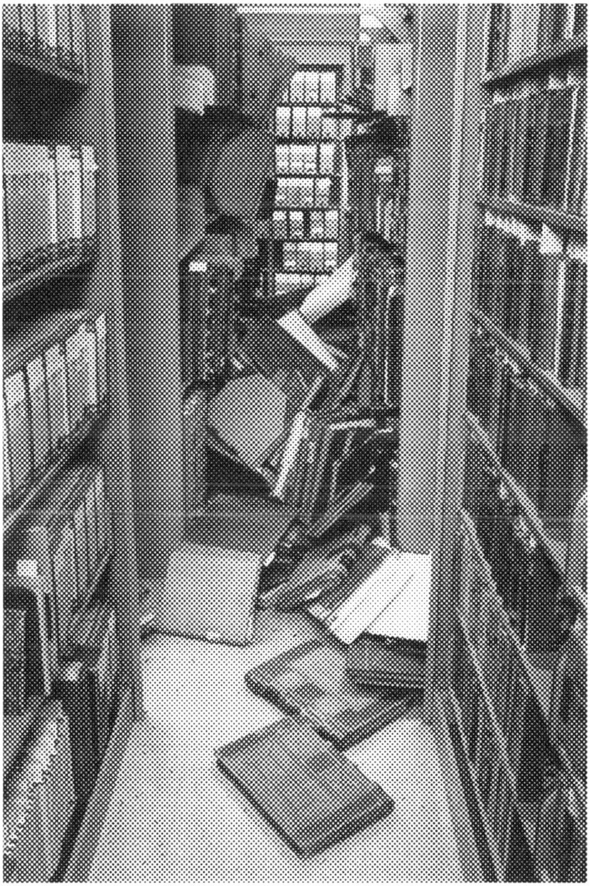

Big Bear, USA

28 June 1992

University of California Riverside Library

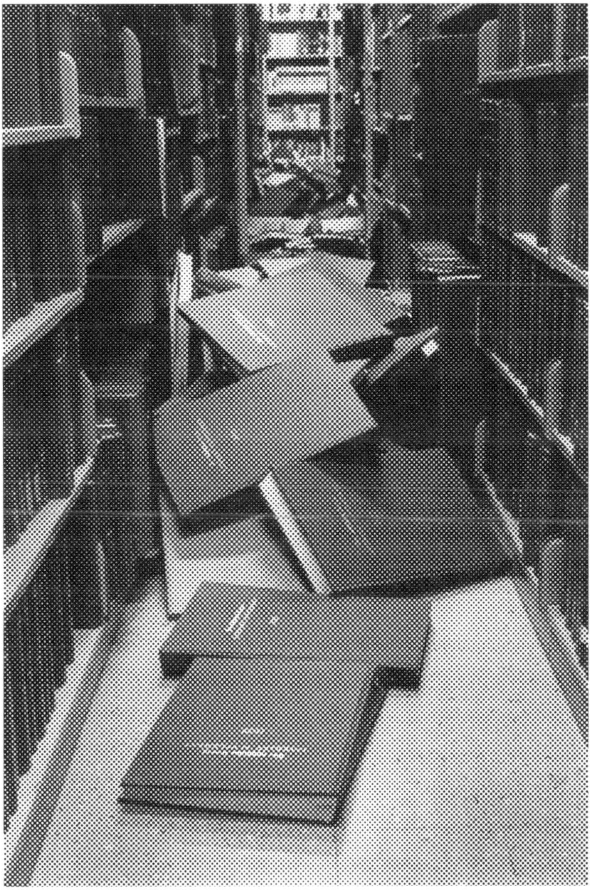

Big Bear, USA

28 June 1992

University of California Riverside Library

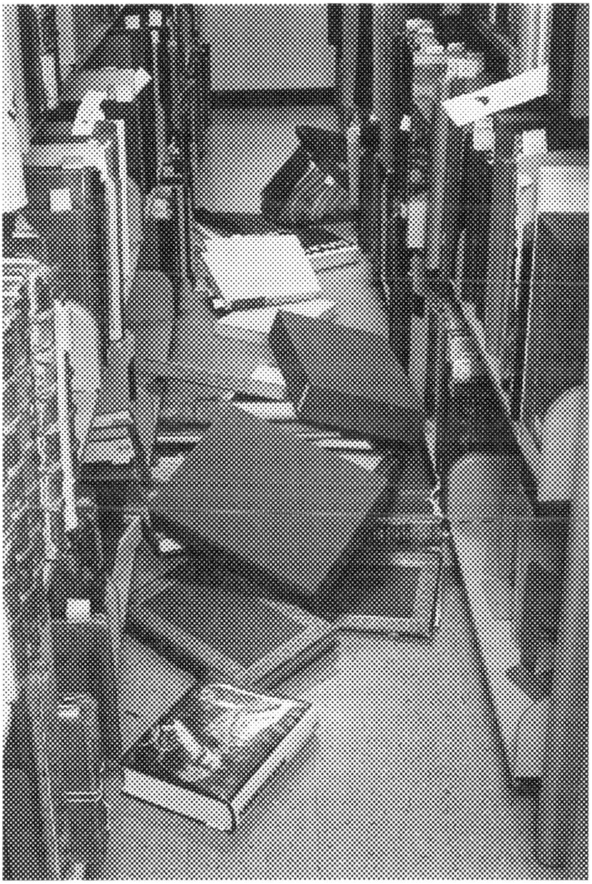

Big Bear, USA

28 June 1992

Santa Monica High School Library

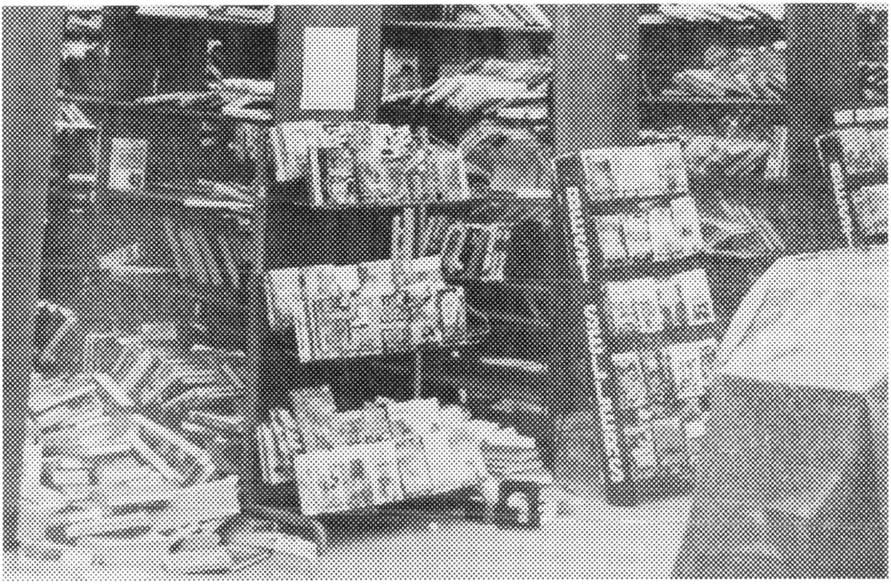

Northridge, USA

17 January 1994

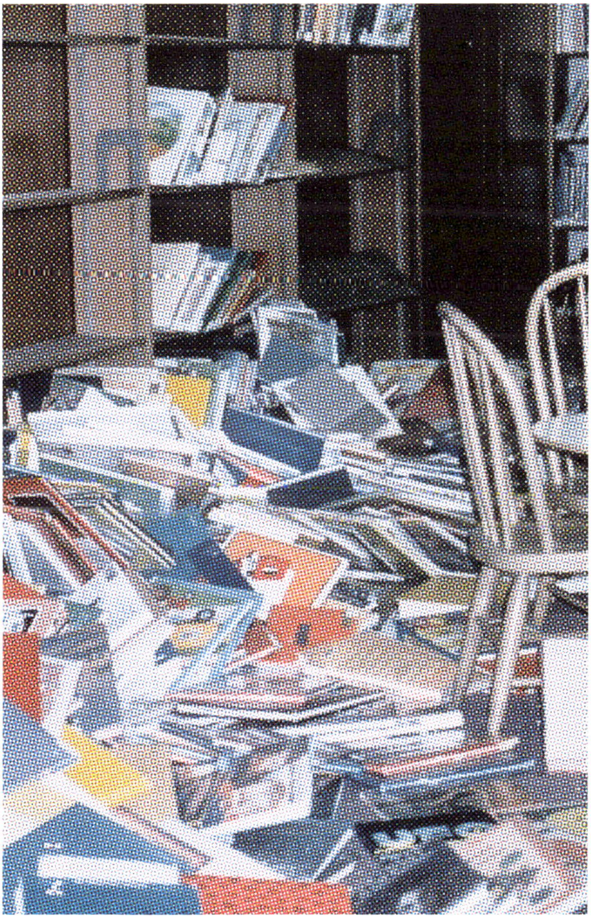

Northridge, USA

17 January 1994

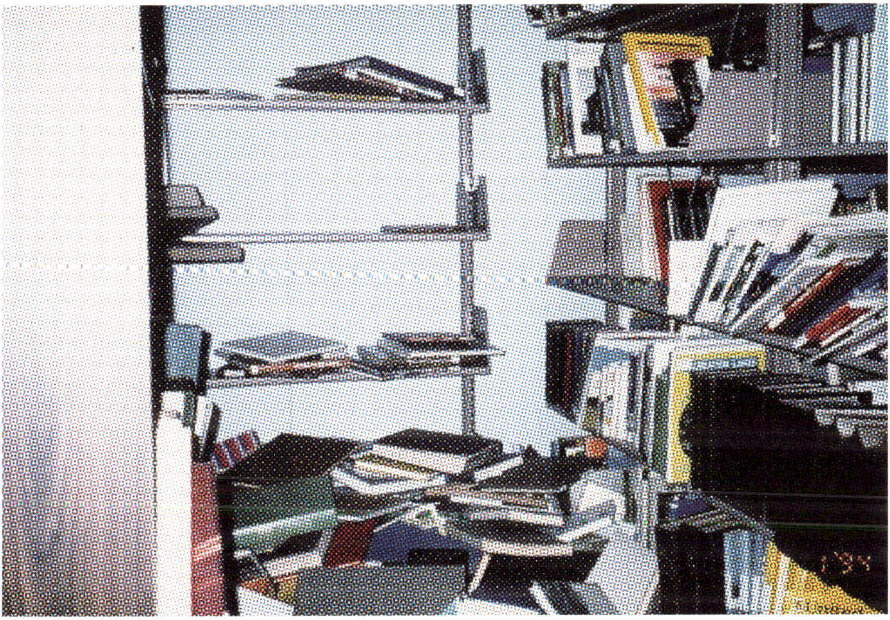

<u>Northridge</u>, USA

17 January 1994

Santa Monica Public Library

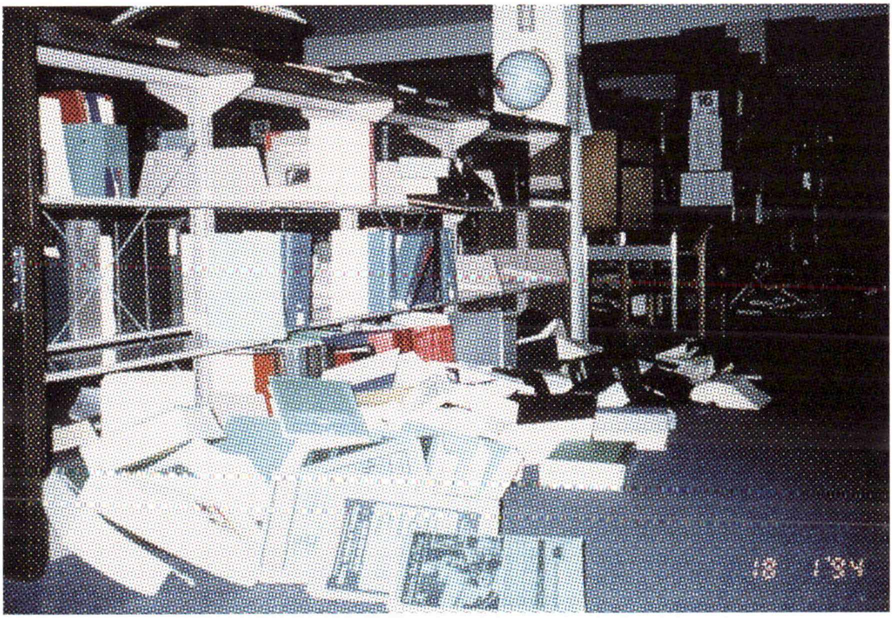

Northridge, USA

17 January 1994

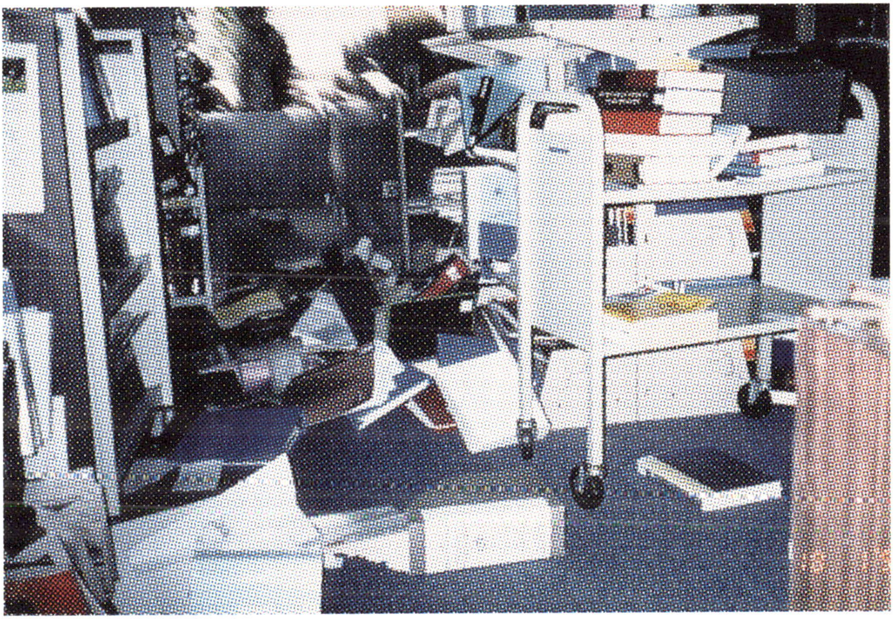

Northridge, USA

17 January 1994

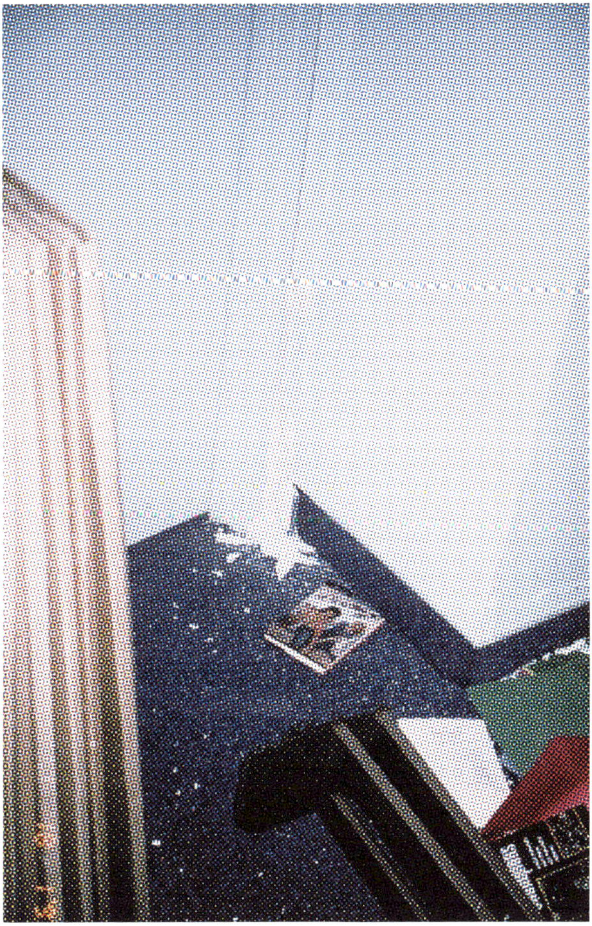

Northridge, USA

17 January 1994

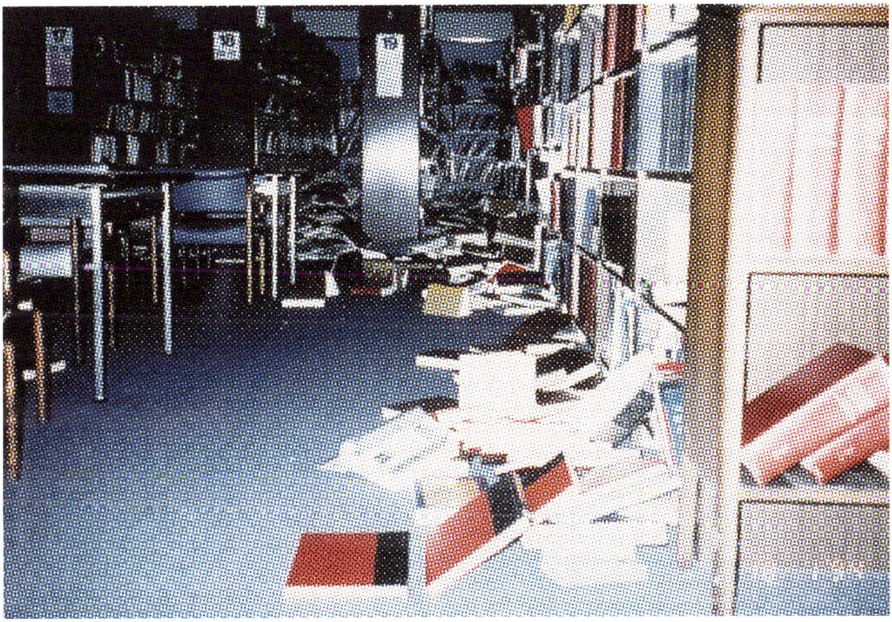

Northridge, USA

17 January 1994

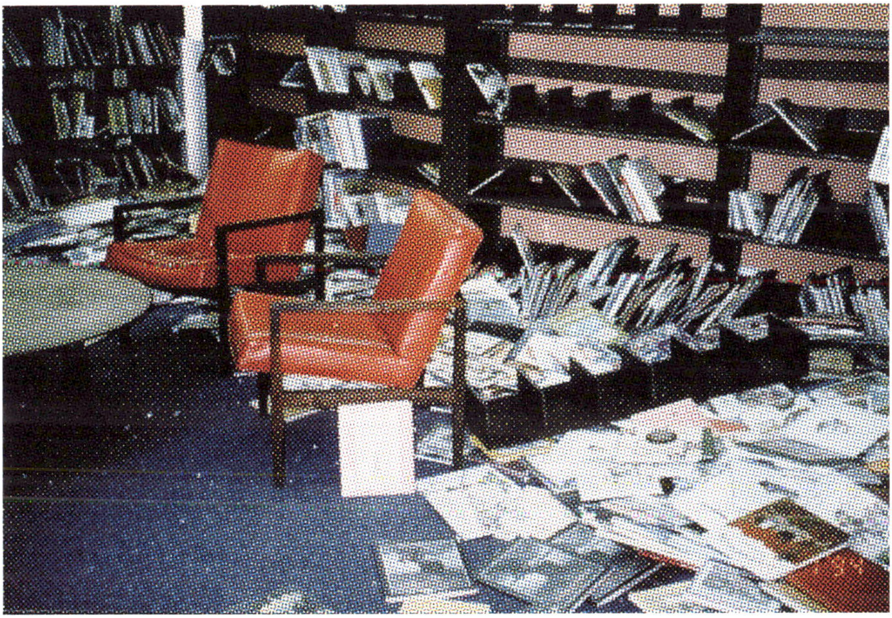

Northridge, USA

17 January 1994

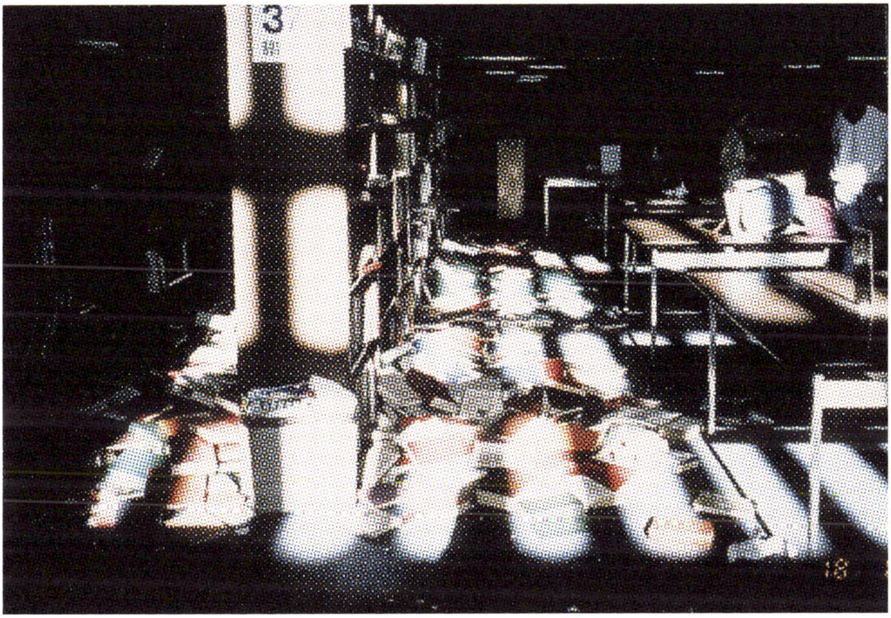

Northridge, USA

17 January 1994

Santa Monica Public Library

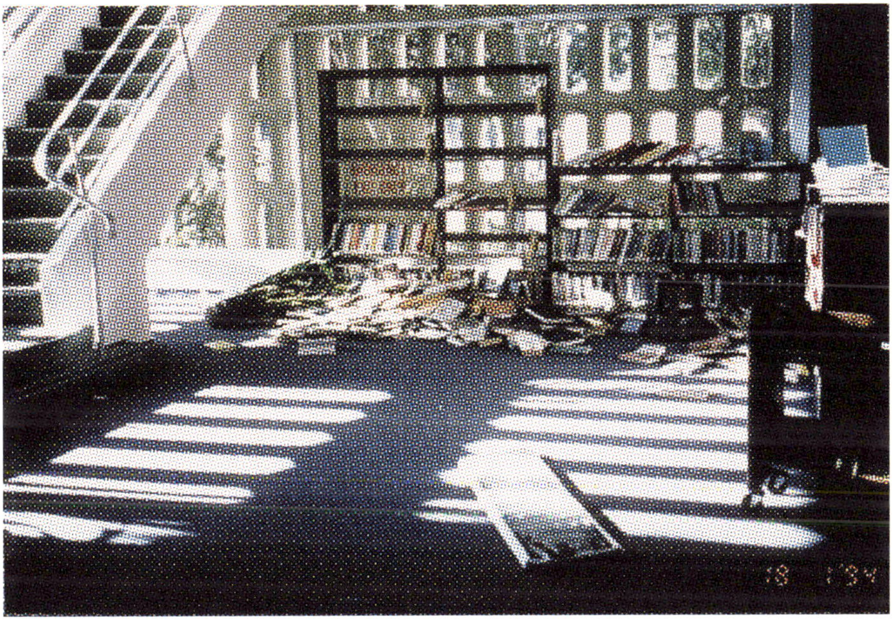

Northridge, USA

17 January 1994

Santa Monica Public Library

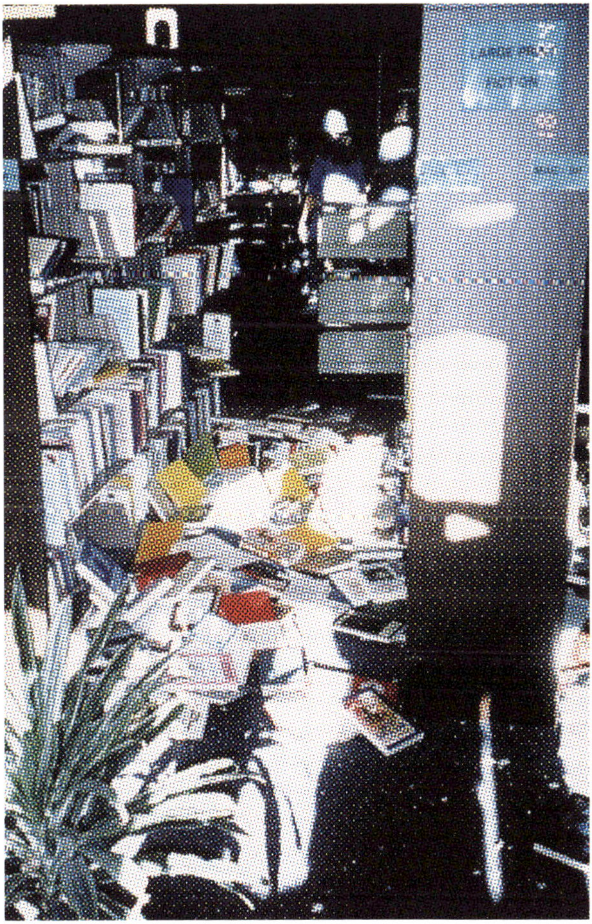

Northridge, USA

17 January 1994

Santa Monica Public Library

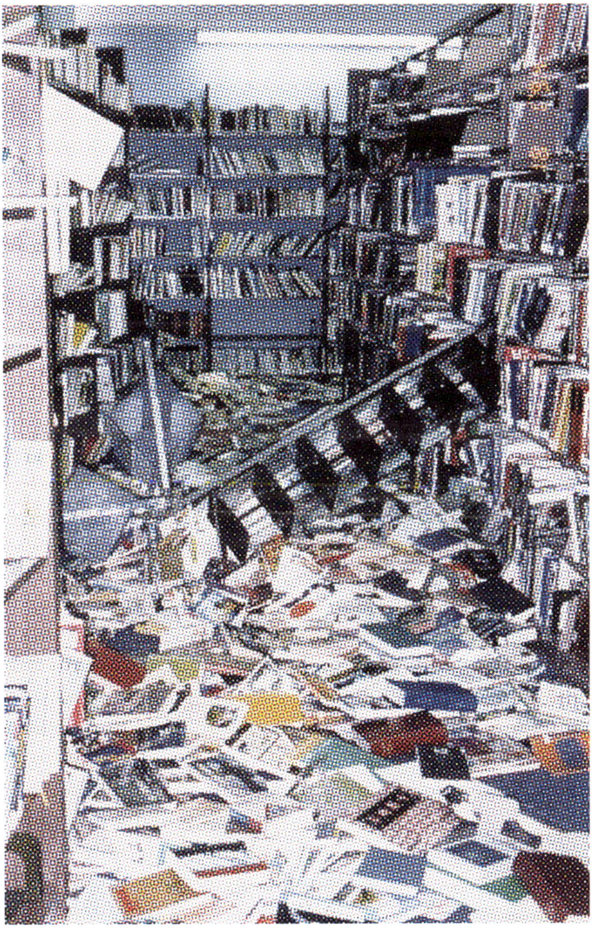

Northridge, USA

17 January 1994

Santa Monica Public Library

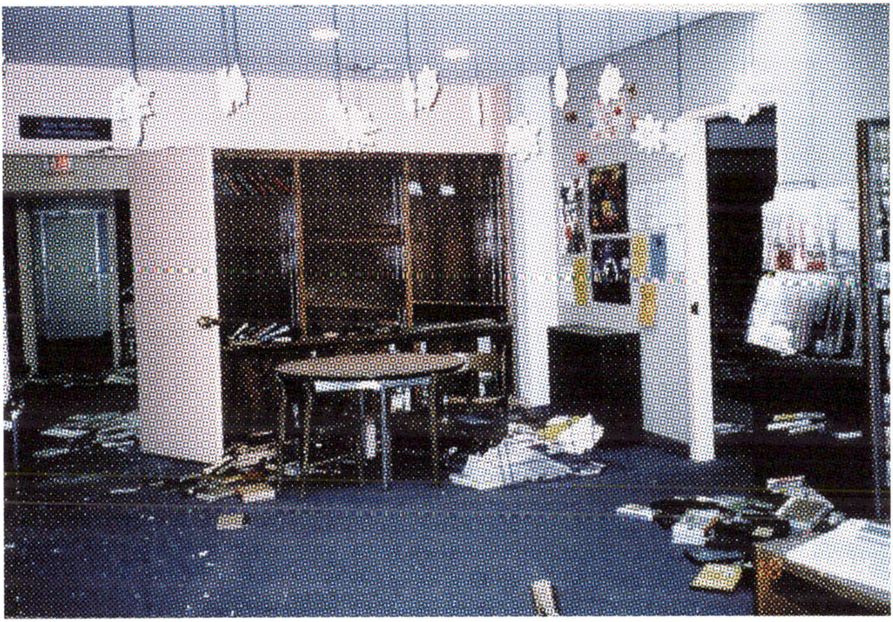

Northridge, USA

17 January 1994

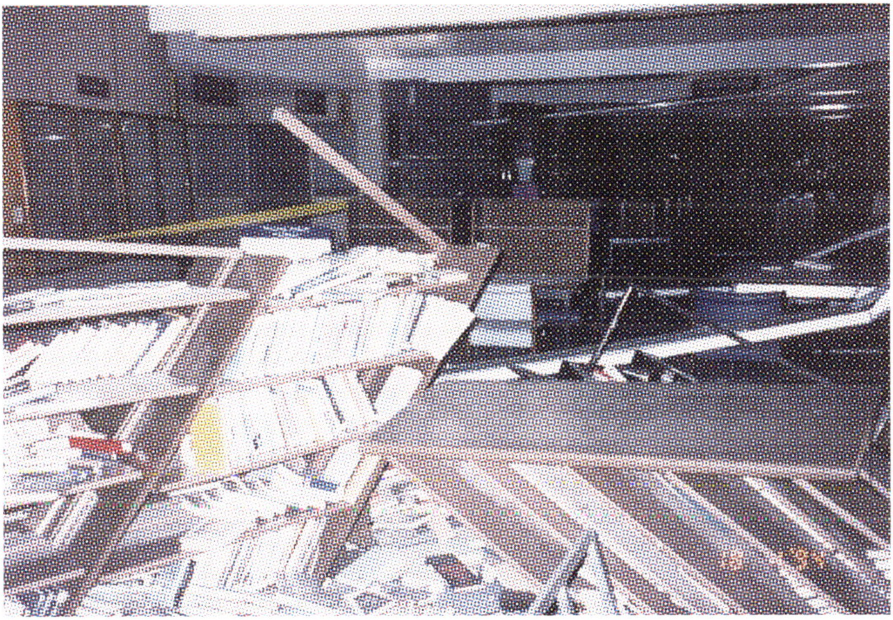

Northridge, USA

17 January 1994

Santa Monica Public Library

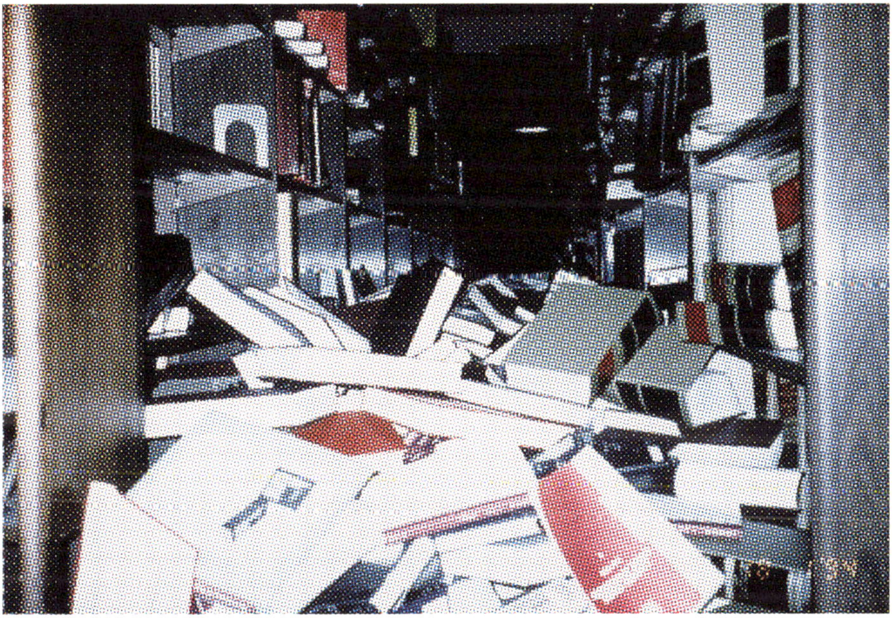

Northridge, USA

17 January 1994

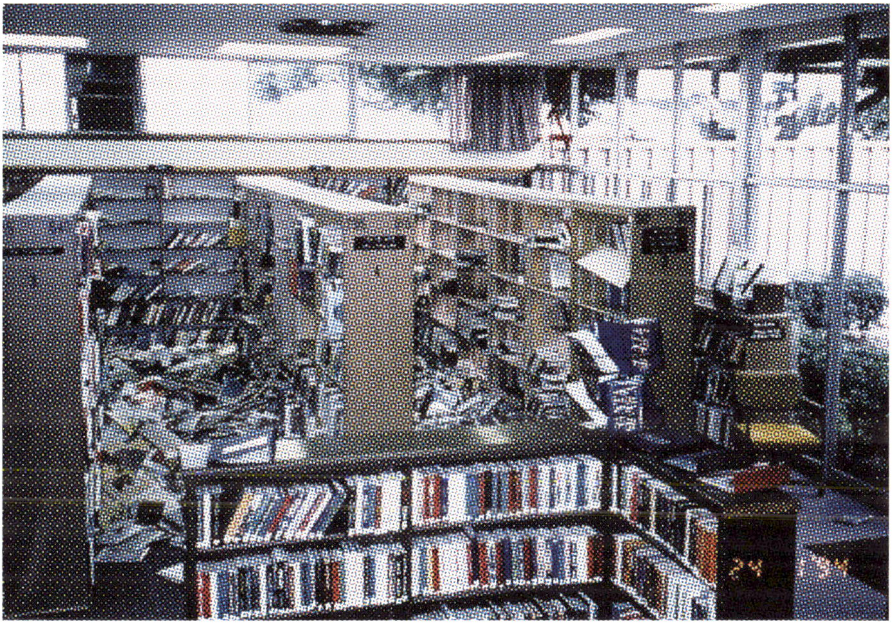

Northridge, USA

17 January 1994

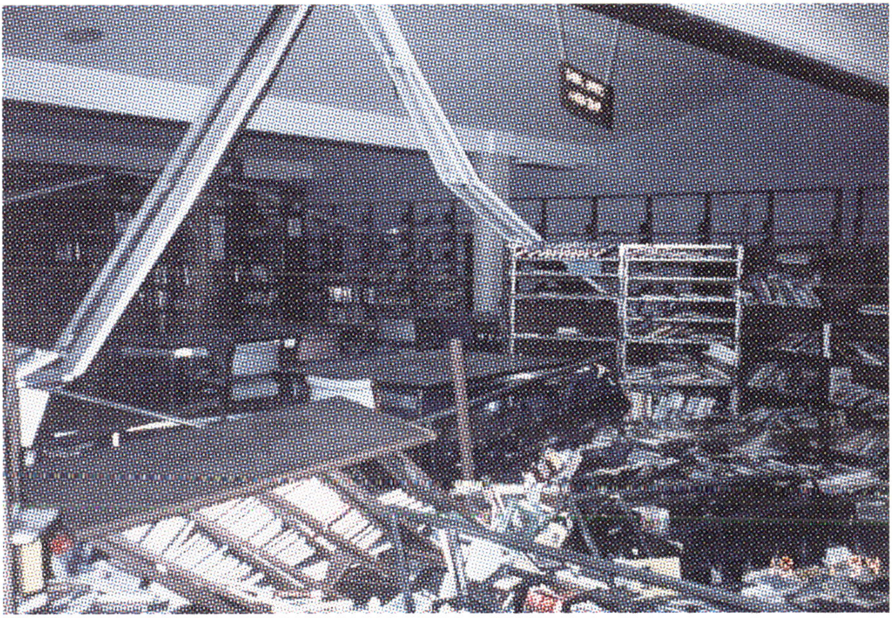

Northridge, USA

17 January 1994

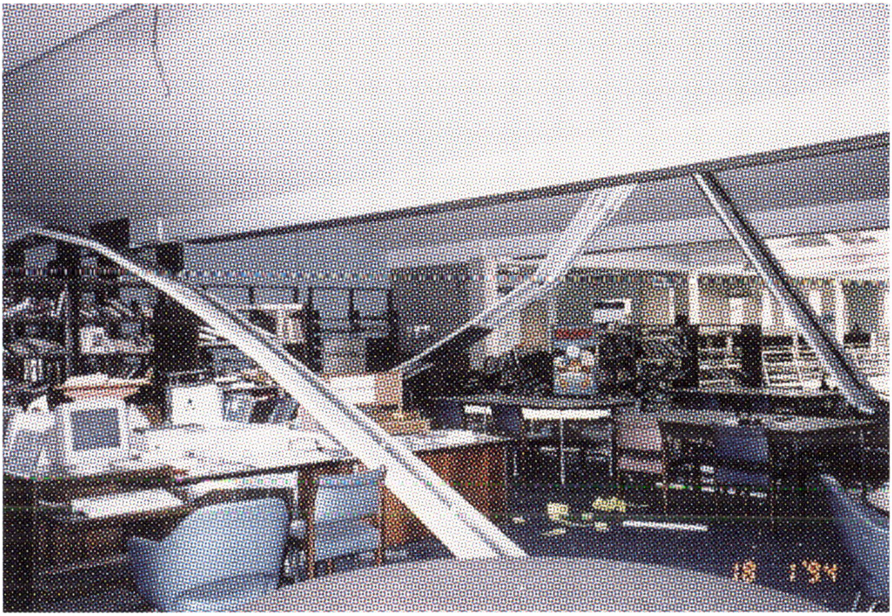

Northridge, USA

17 January 1994

Santa Monica Public Library

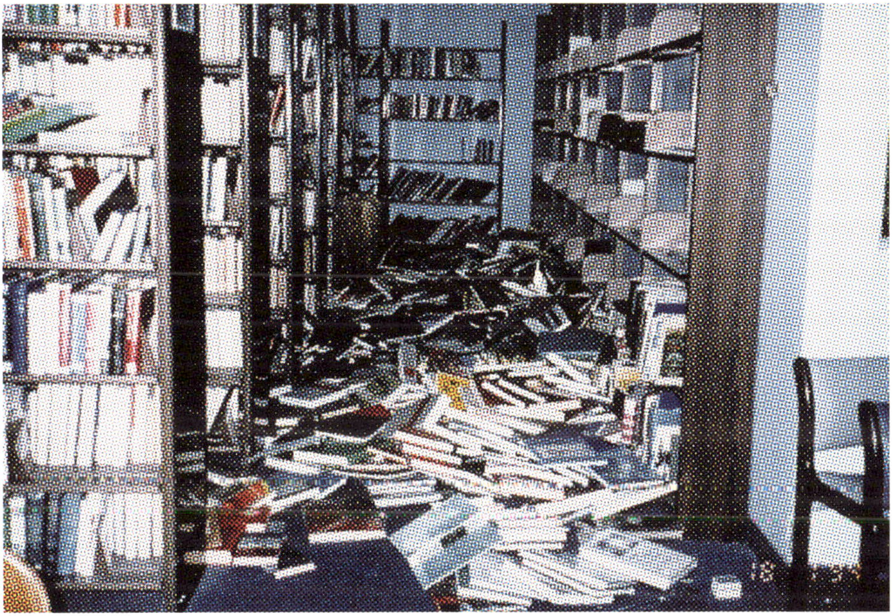

Northridge, USA

17 January 1994

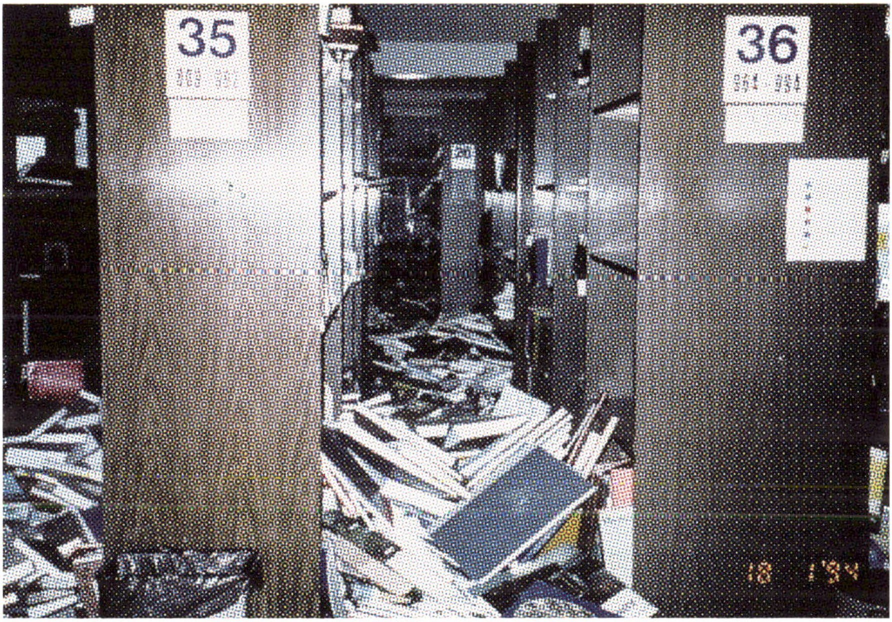
35
36

Northridge, USA

17 January 1994

Santa Monica Public Library

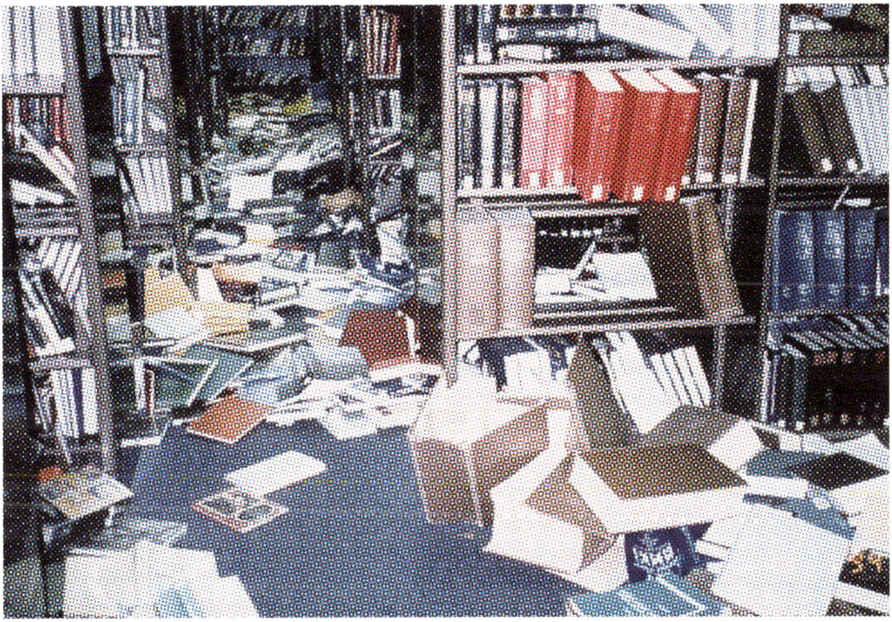

Northridge, USA

17 January 1994

CSU Northridge Oviatt Library

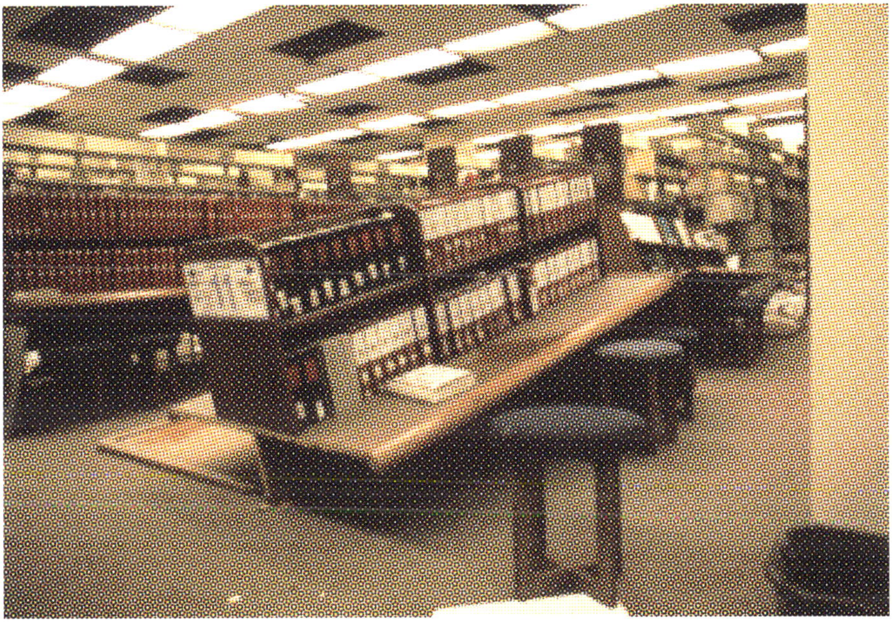

Northridge, USA

17 January 1994

CSU Northridge Oviatt Library

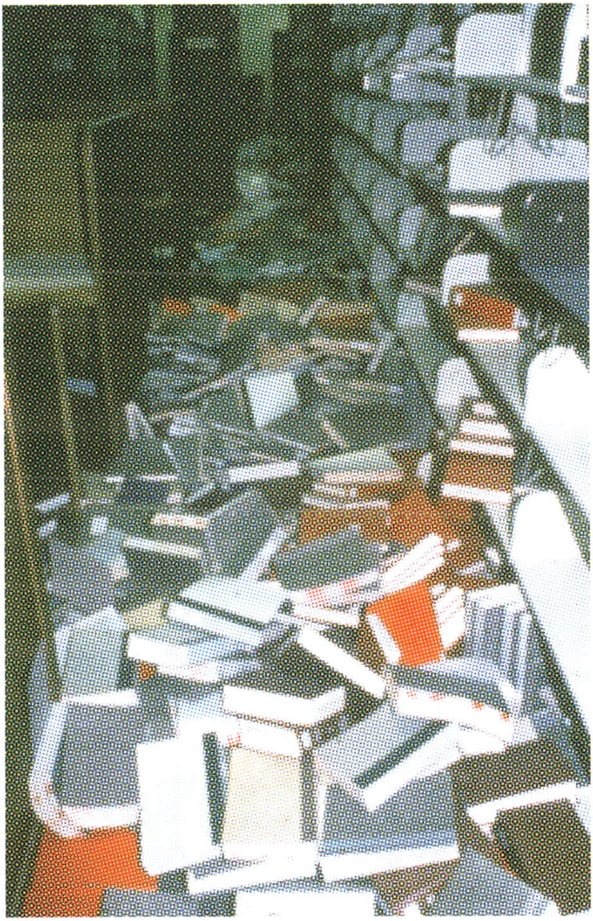

Northridge, USA

17 January 1994

Kobe University Library

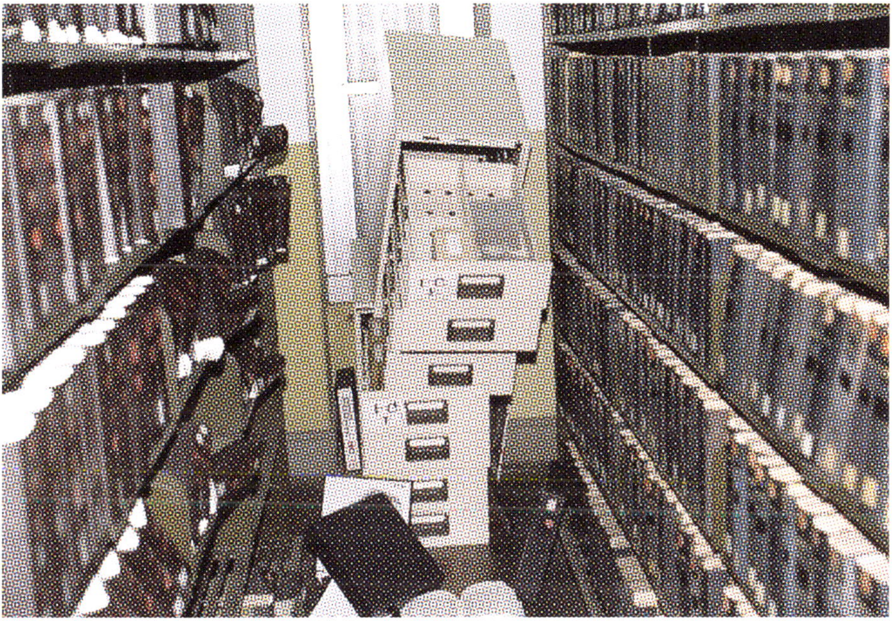

Hanshin-Awaji, Japan

17 January 1995

Kobe University Library

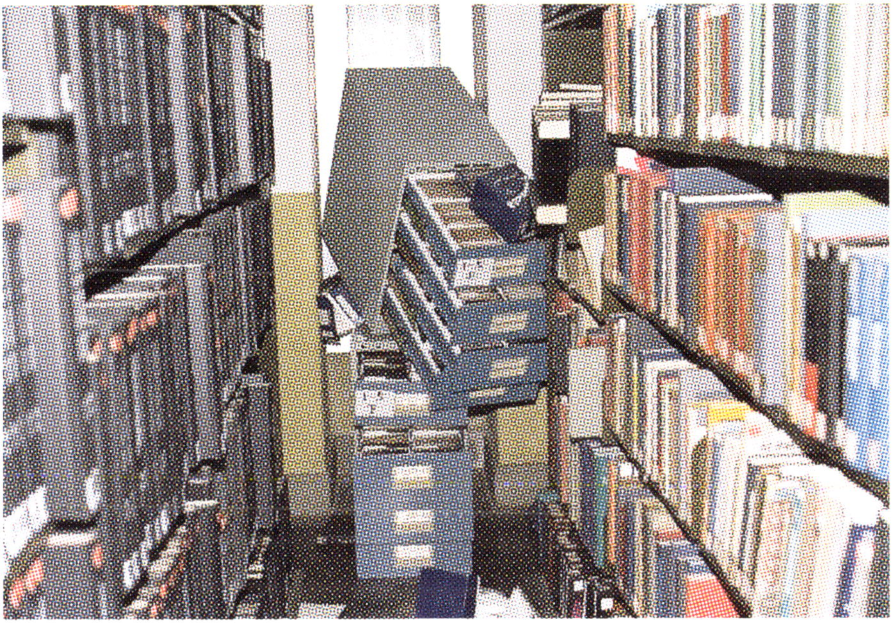

Hanshin-Awaji, Japan

17 January 1995

Kobe University Library

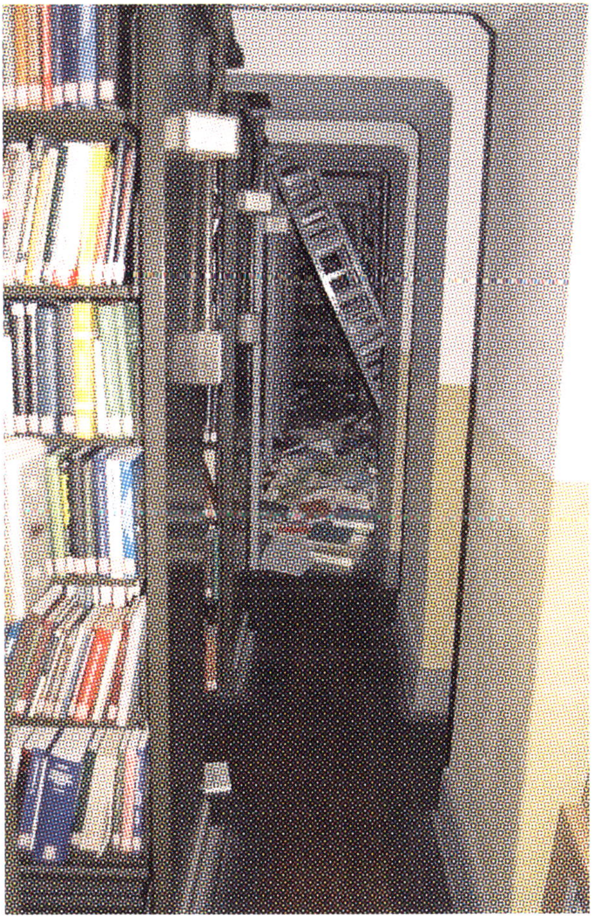

Hanshin-Awaji, Japan

17 January 1995

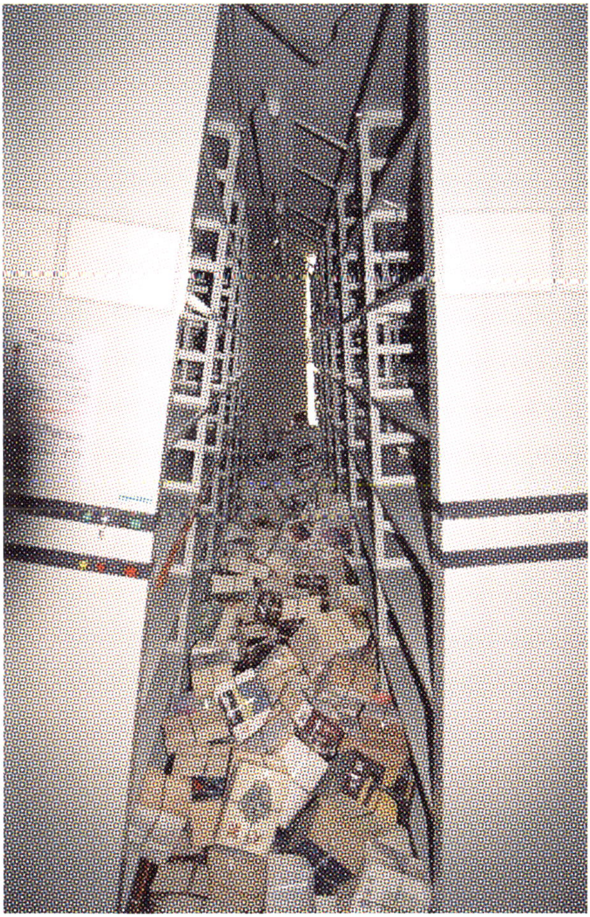

Hanshin-Awaji, Japan

17 January 1995

Kobe University Library

Hanshin-Awaji, Japan

17 January 1995

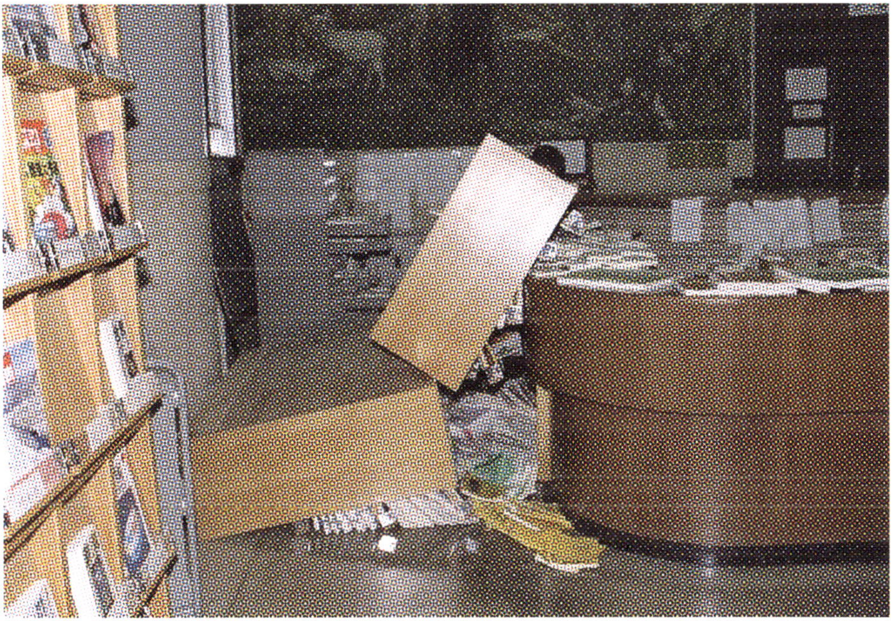

Hanshin-Awaji, Japan

17 January 1995

Kobe University Library

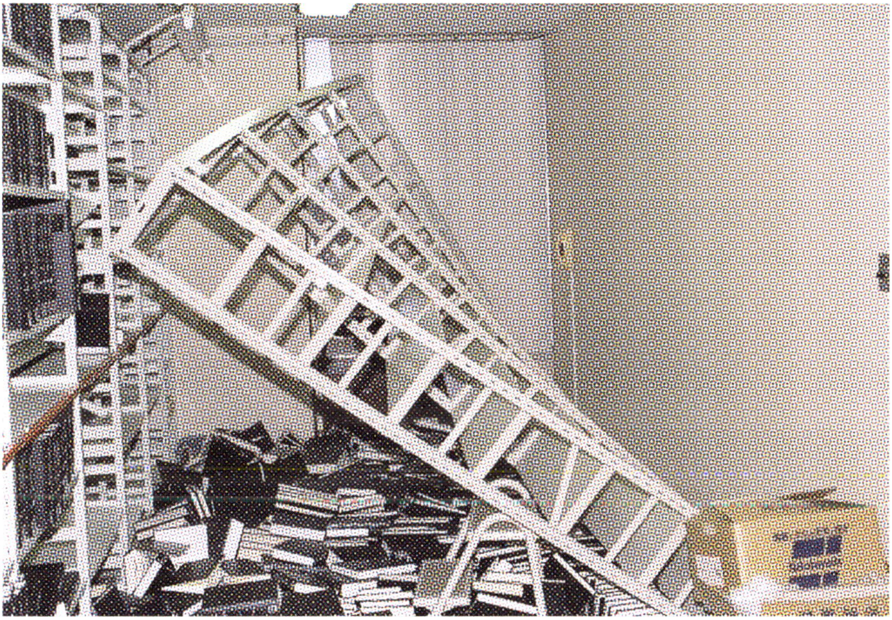

Hanshin-Awaji, Japan

17 January 1995

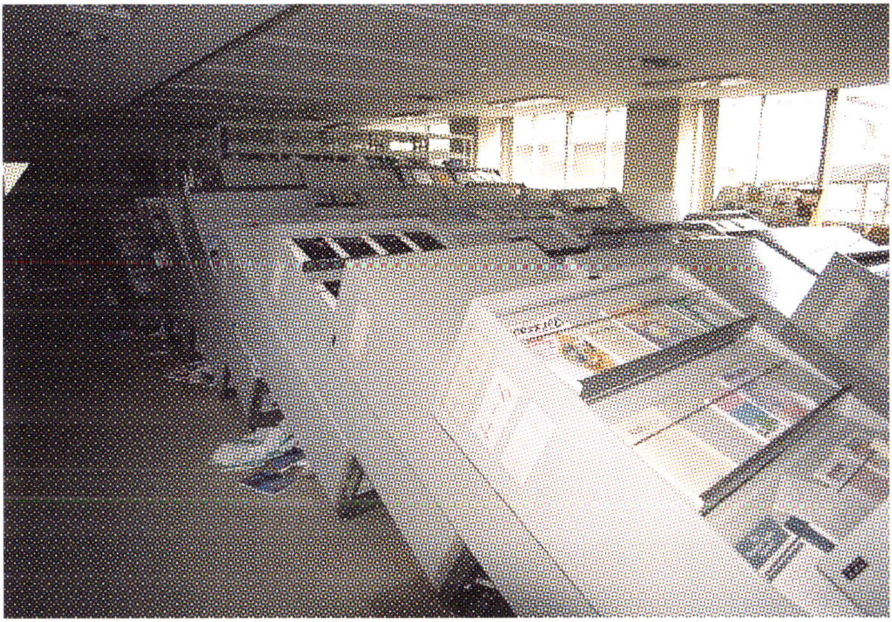

Hanshin-Awaji, Japan

17 January 1995

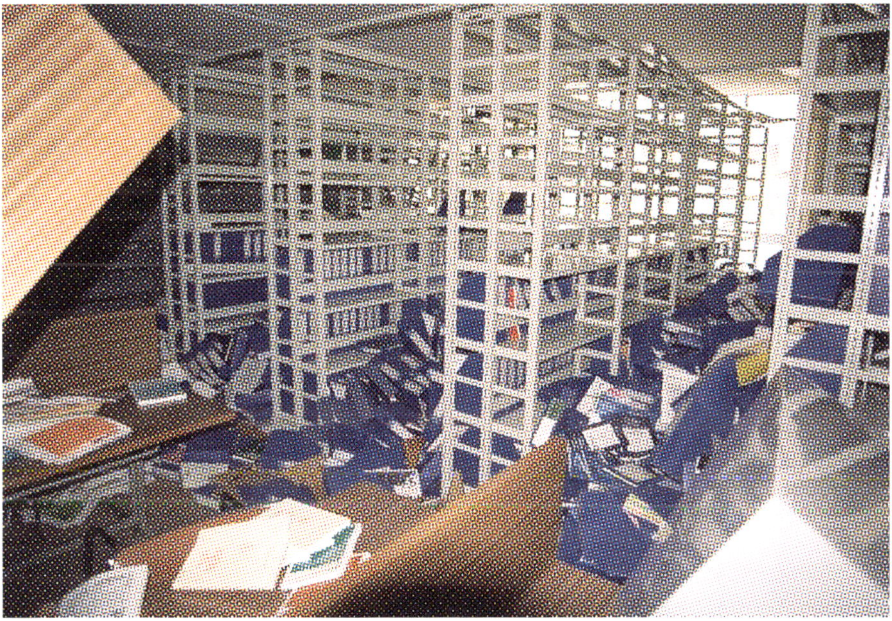

Hanshin-Awaji, Japan

17 January 1995

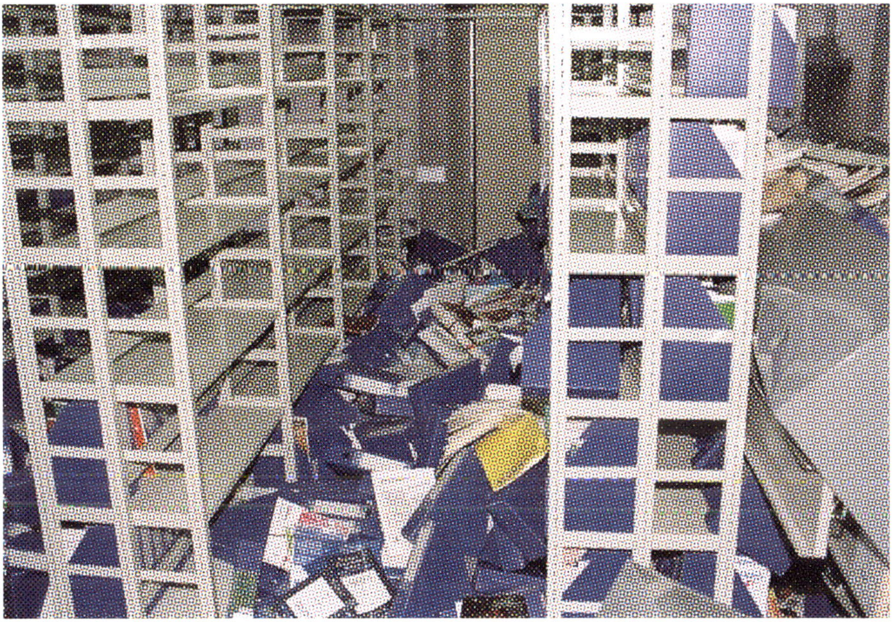

Hanshin-Awaji, Japan

17 January 1995

Kobe University Library

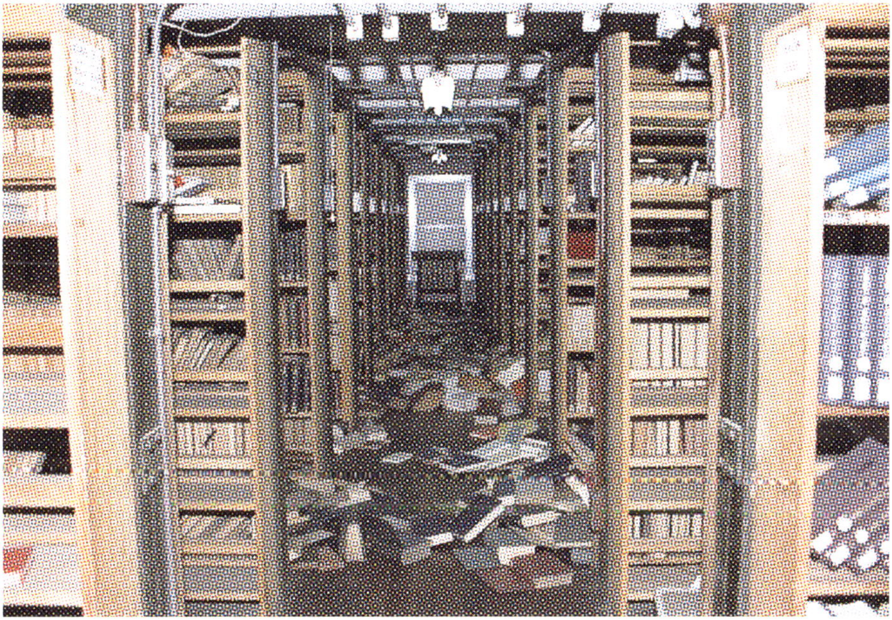

<u>Hanshin-Awaji</u>, Japan

17 January 1995

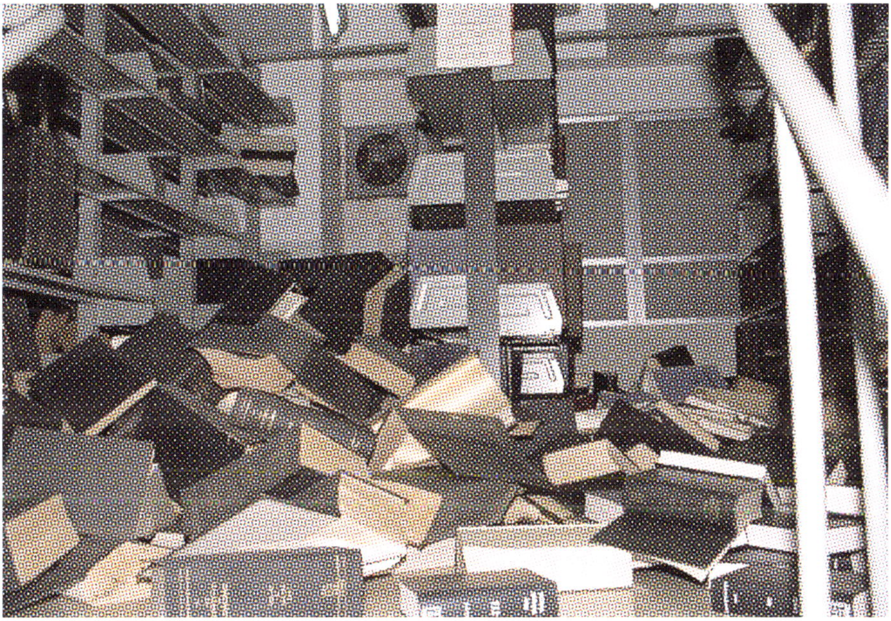

Hanshin-Awaji, Japan

17 January 1995

Kobe University Library

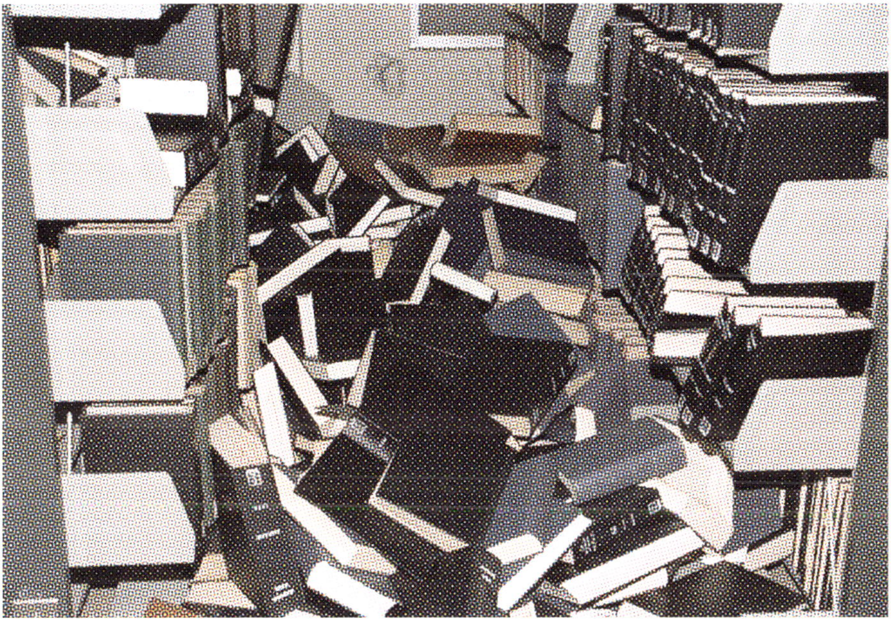

Hanshin-Awaji, Japan

17 January 1995

Kobe University Library

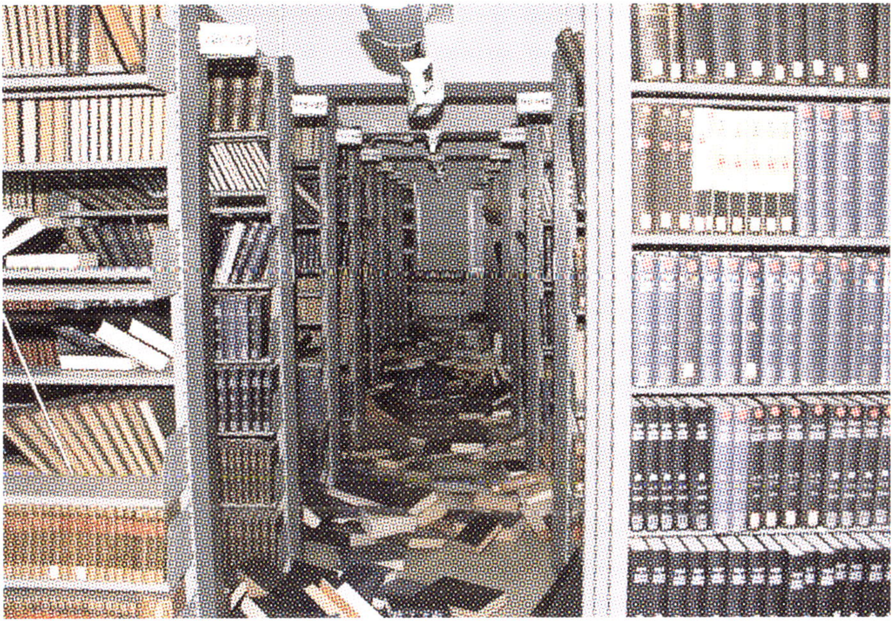

Hanshin-Awaji, Japan

17 January 1995

Kobe University Library

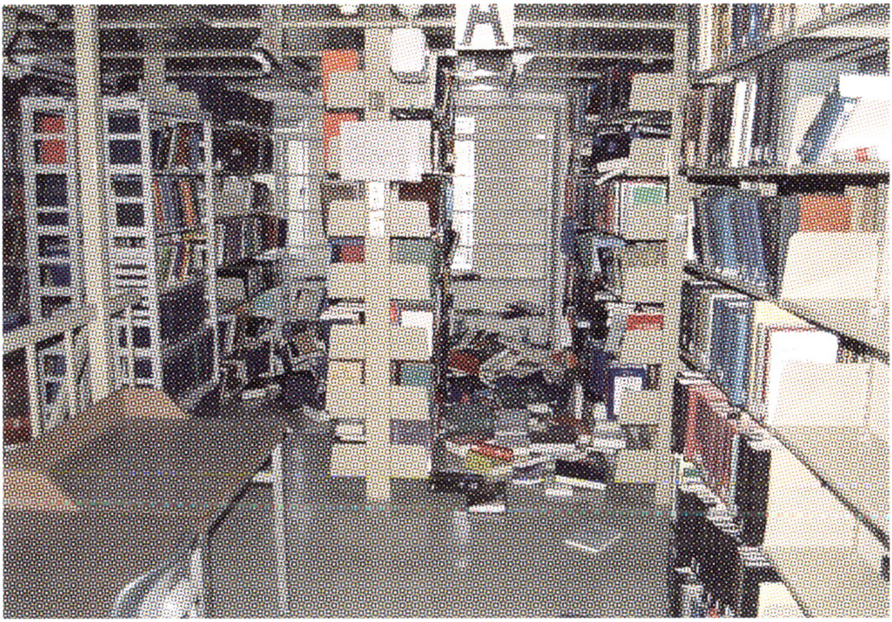

Hanshin-Awaji, Japan

17 January 1995

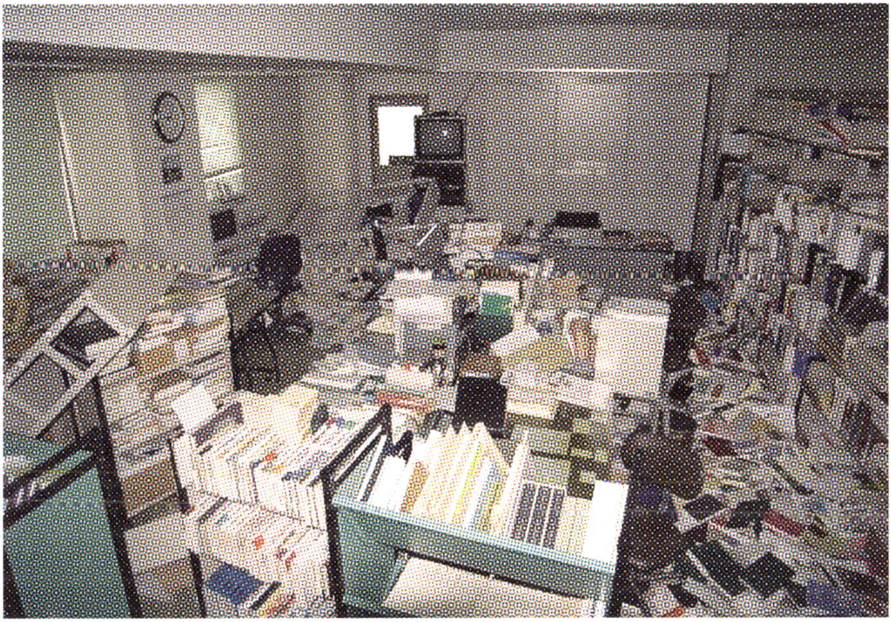

Hanshin-Awaji, Japan

17 January 1995

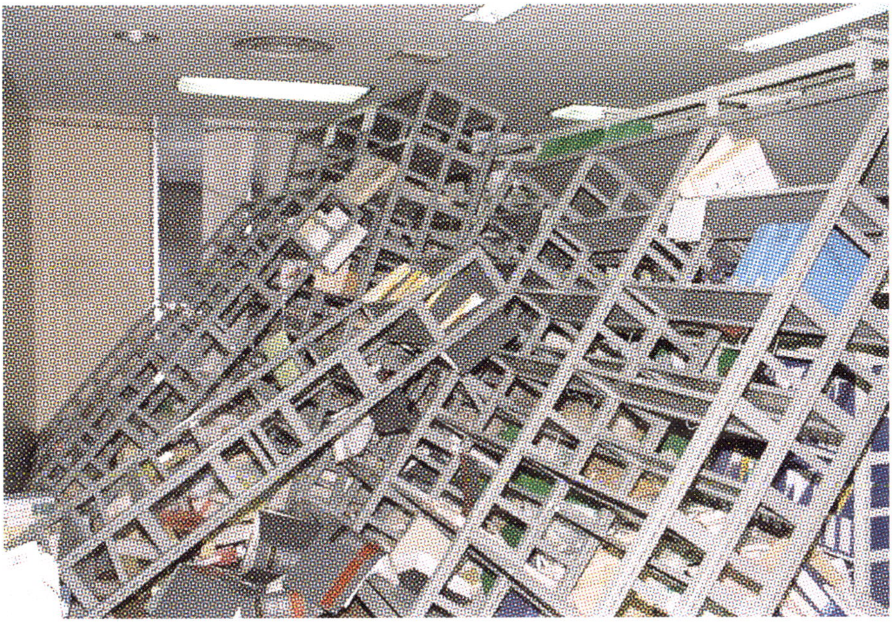

Hanshin-Awaji, Japan

17 January 1995

Seattle University School of Law Library

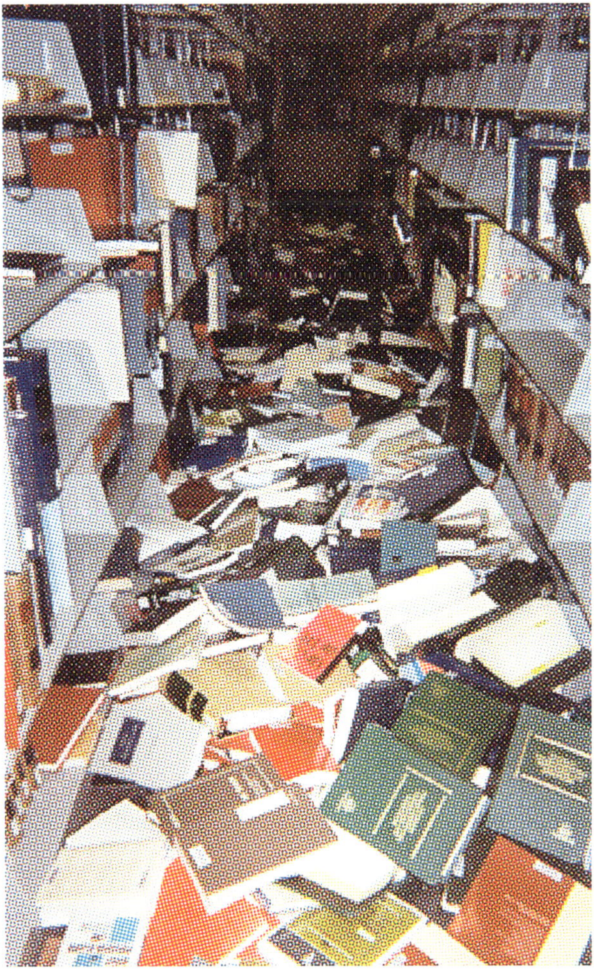

Nisqually, USA

28 February 2001

Seattle University School of Law Library

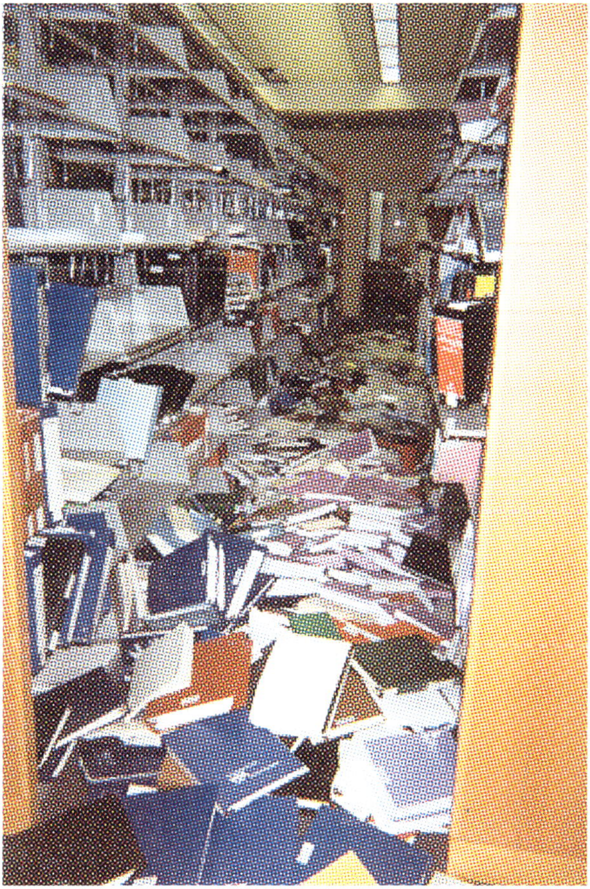

<u>Nisqually</u>, USA

28 February 2001

Seattle University School of Law Library

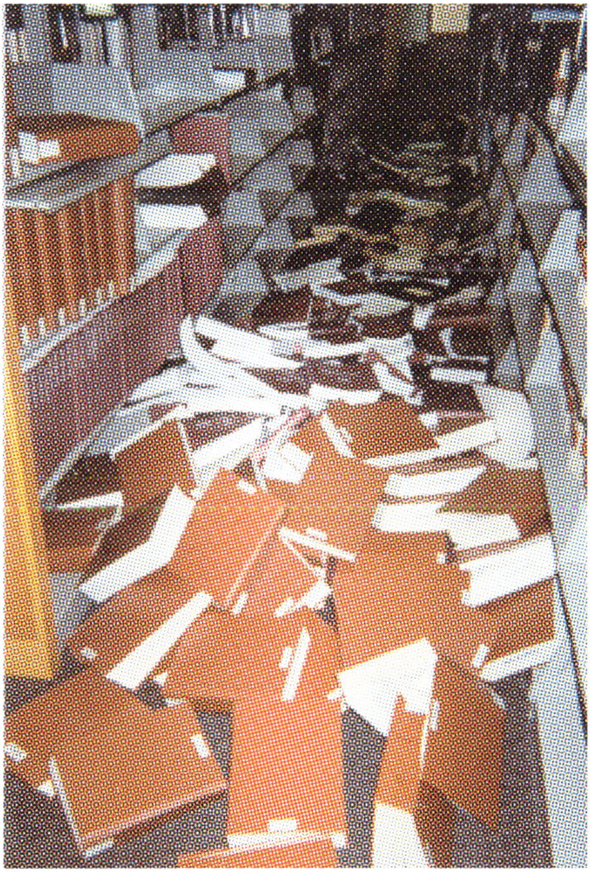

Nisqually, USA

28 February 2001

Seattle University School of Law Library

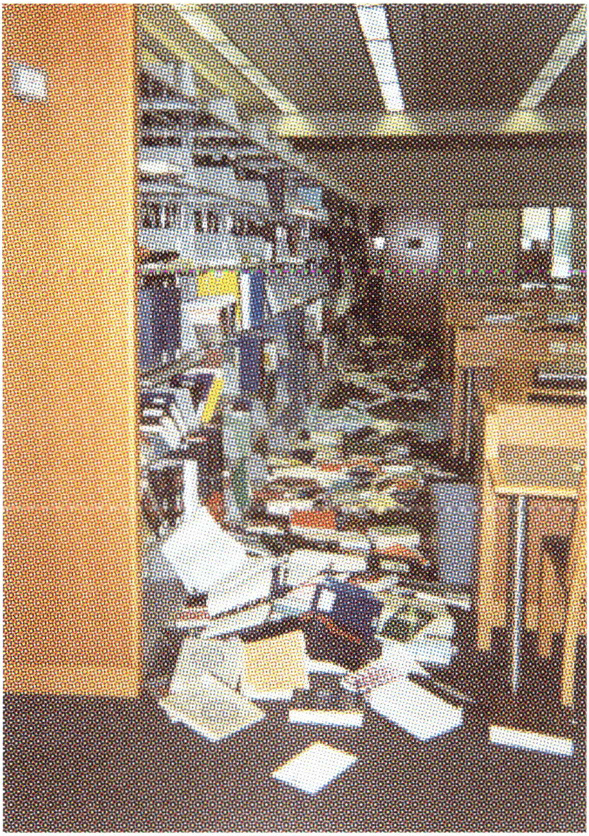

Nisqually, USA

28 February 2001

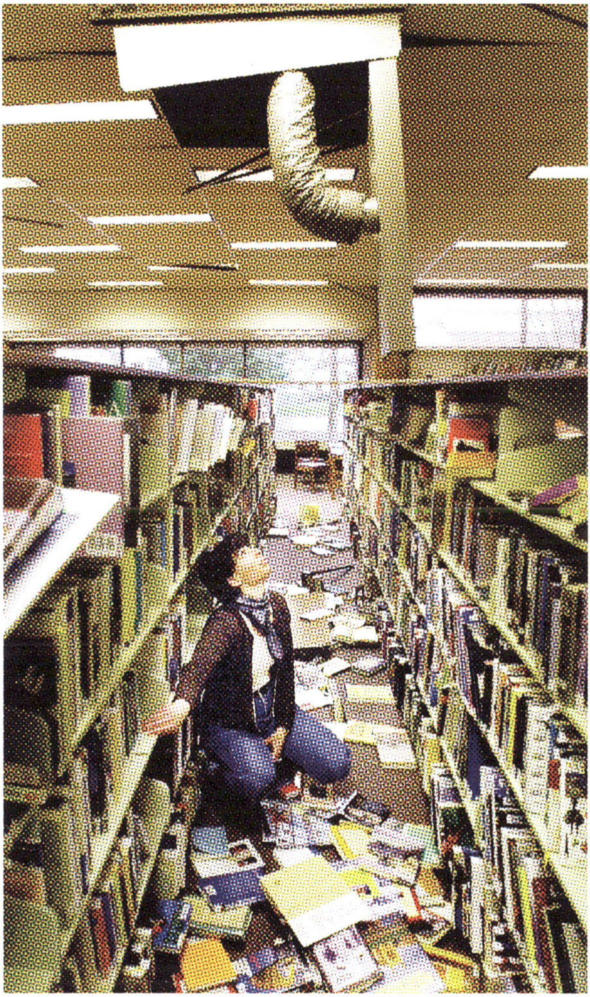

Cheryl Heywood, community librarian for the Olympia Timberland Library, looks over earthquake damage at the downtown library on Thursday. "The library is closed until further notice," she said.

University of Washington Oceanography Library

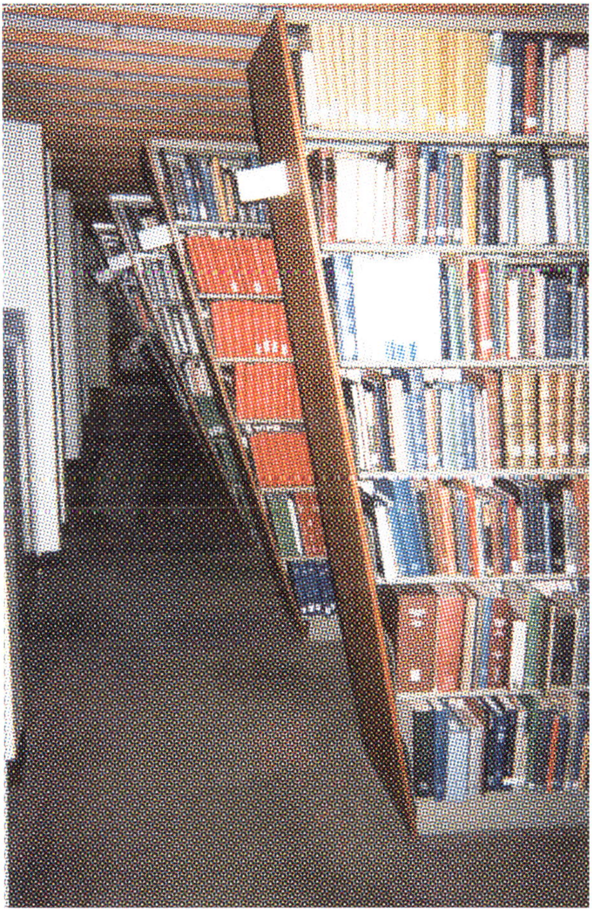

Nisqually, USA

28 February 2001

University of Washington Oceanography Library

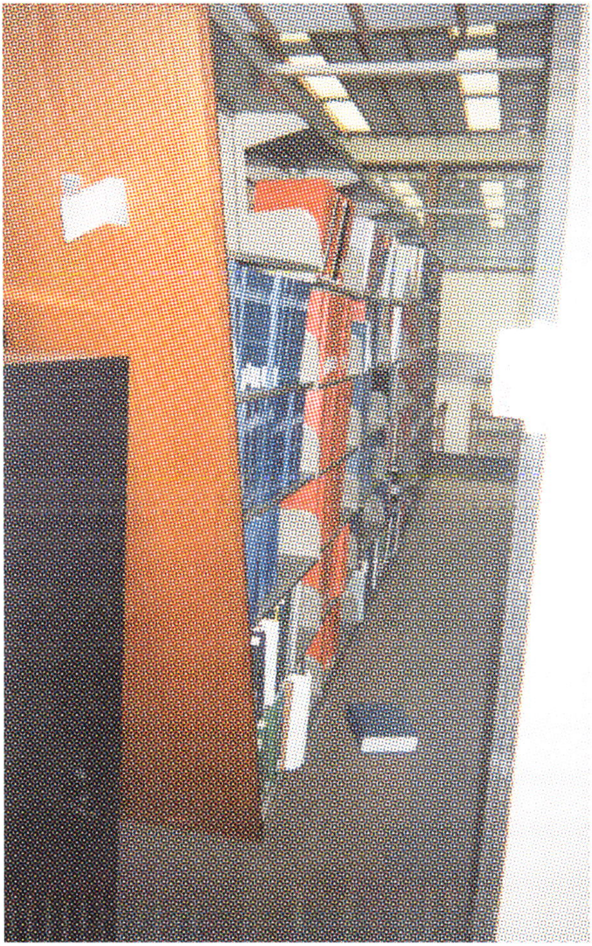

Nisqually, USA

28 February 2001

University of Washington Oceanography Library

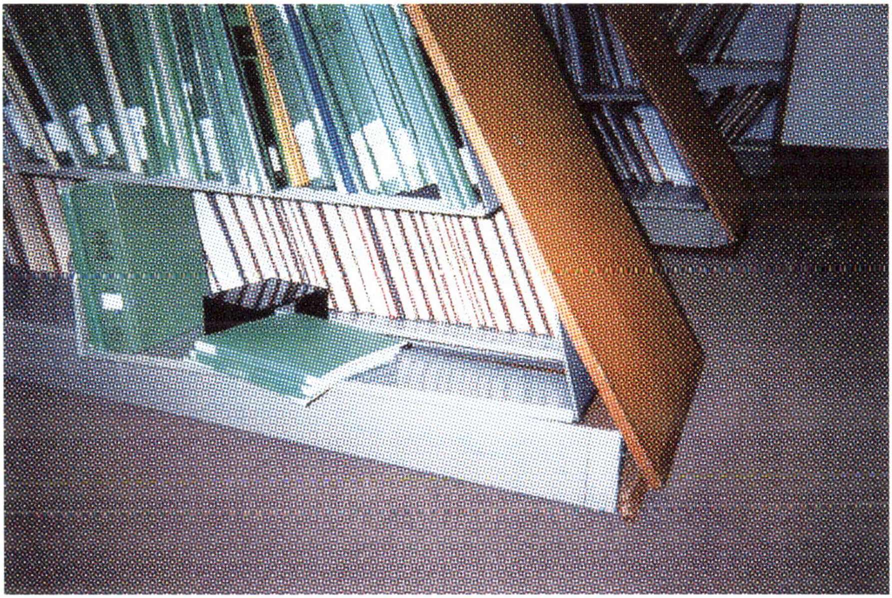

The library had already started weeding books from its selection prior to the quake, having removed 73,000 volumes in the past year. Resorting fallen books has proven helpful in the project to remove unneeded books.

University of Washington Physics Library

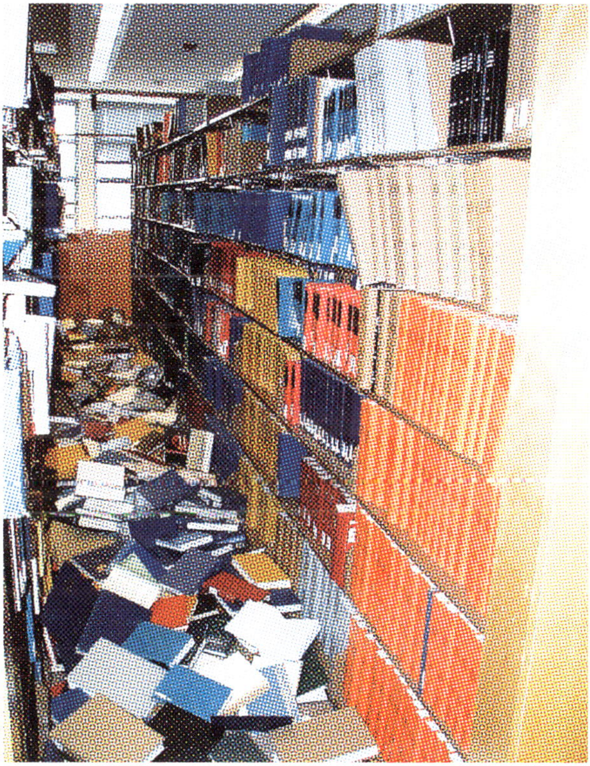

Nisqually, USA

28 February 2001

University of Washington Physics Library

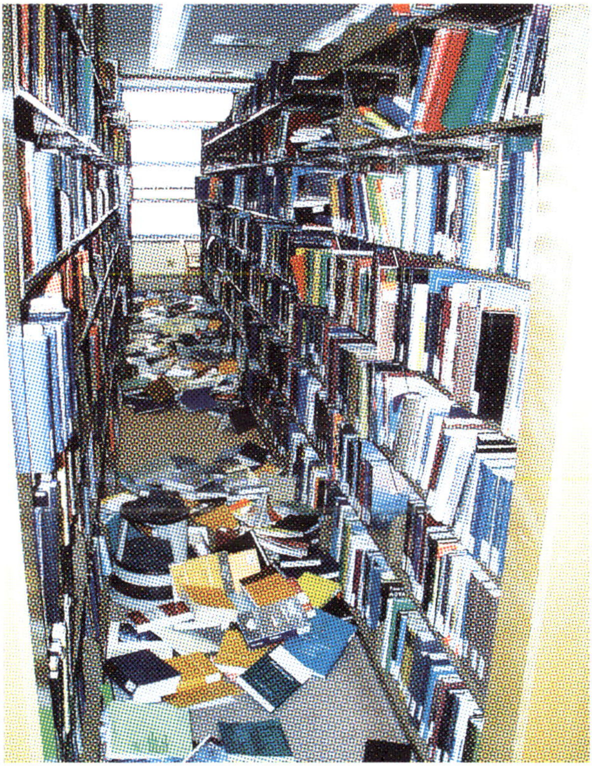

<u>Nisqually</u>, USA

28 February 2001

University of Washington Engineering Library

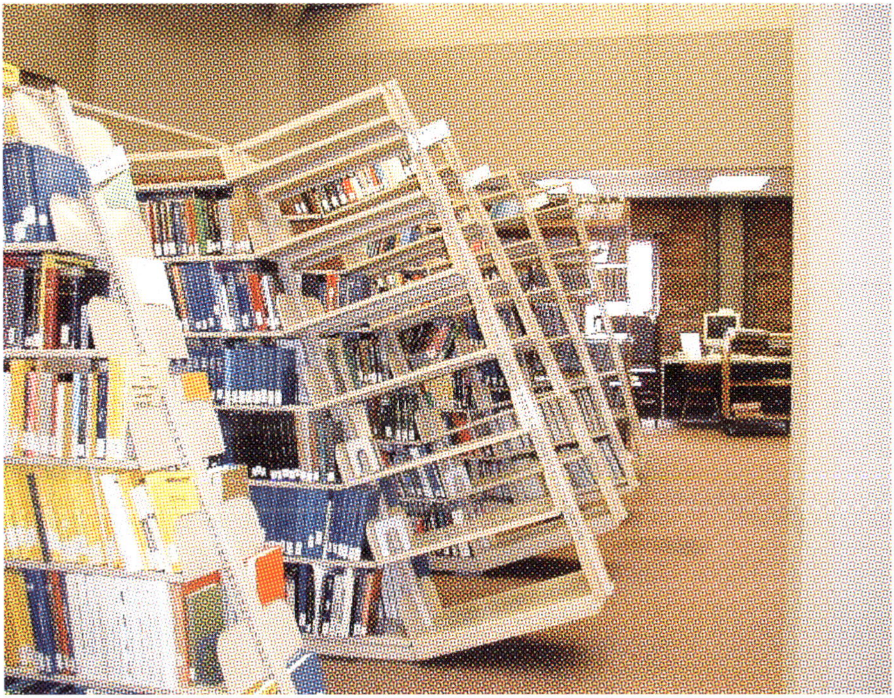

Nisqually, USA

28 February 2001

University of Washington Engineering Library

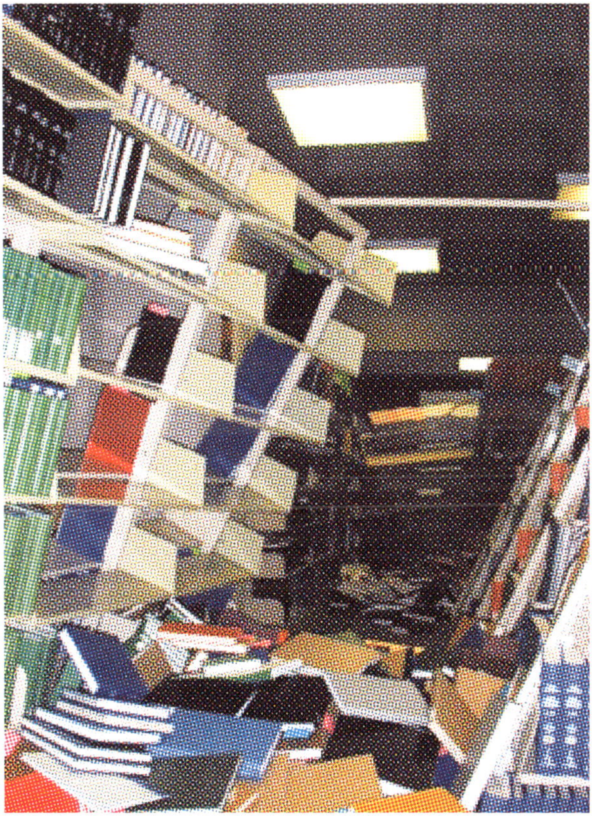

Nisqually, USA

28 February 2001

University of Washington Engineering Library

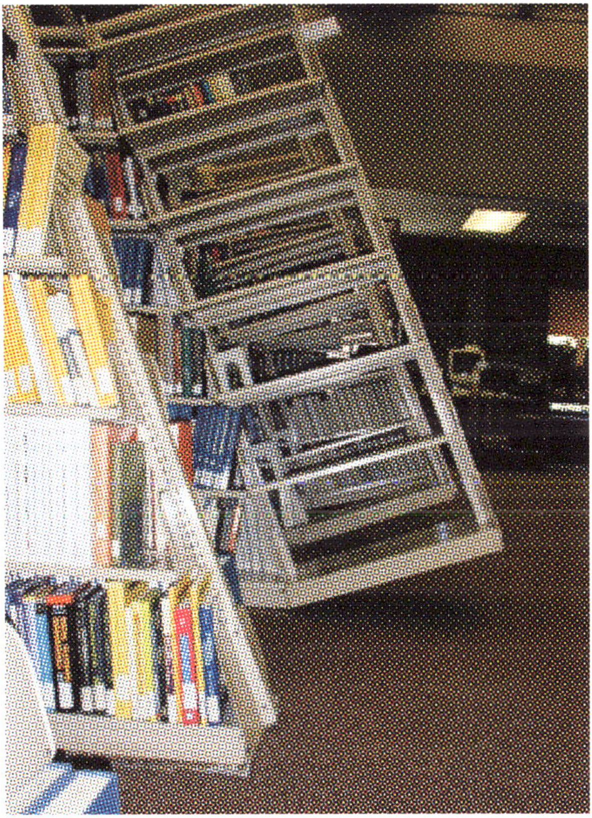

Nisqually, USA

28 February 2001

University of Washington Engineering Library

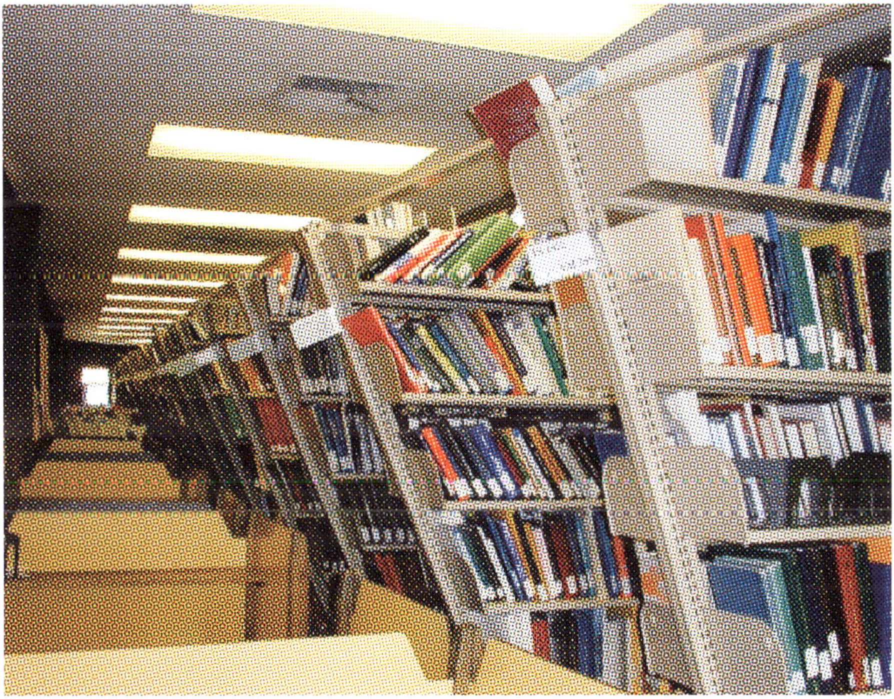

<u>Nisqually</u>, USA

28 February 2001

University of Washington Engineering Library

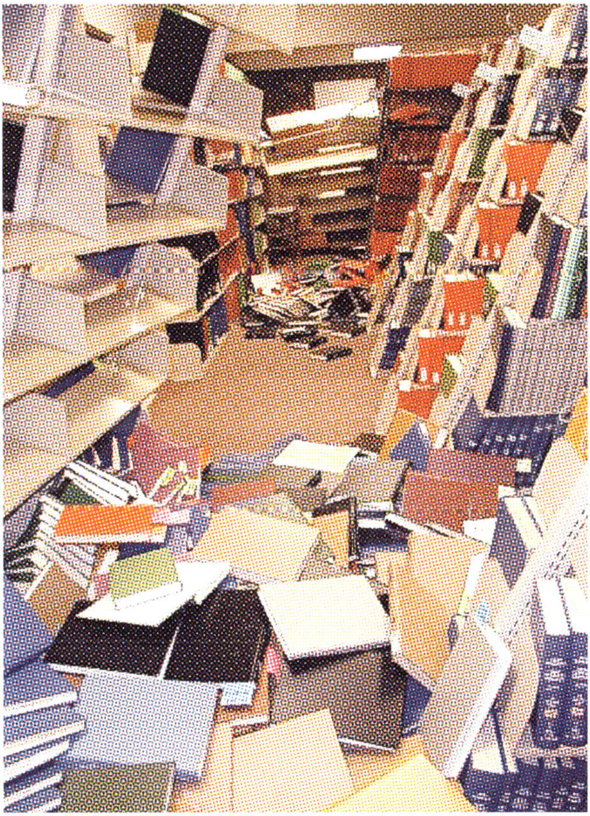

<u>Nisqually</u>, USA

28 February 2001

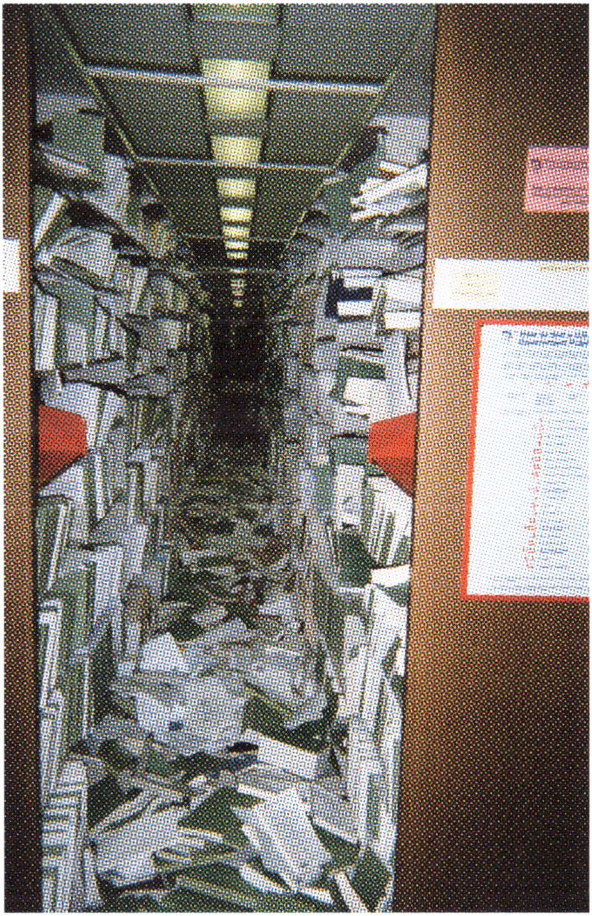

Nisqually, USA

Congressional hearings on the floor.

28 February 2001

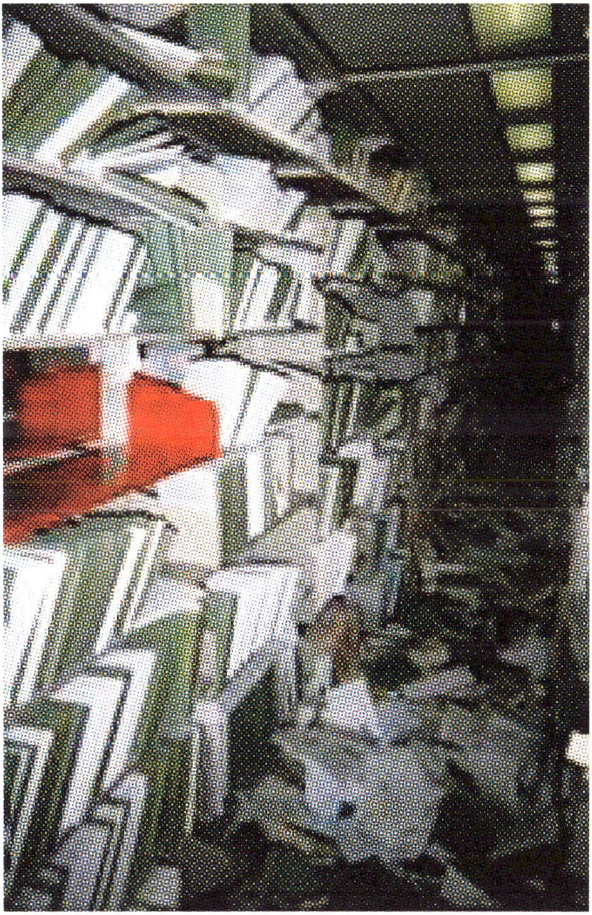

Nisqually, USA

28 February 2001

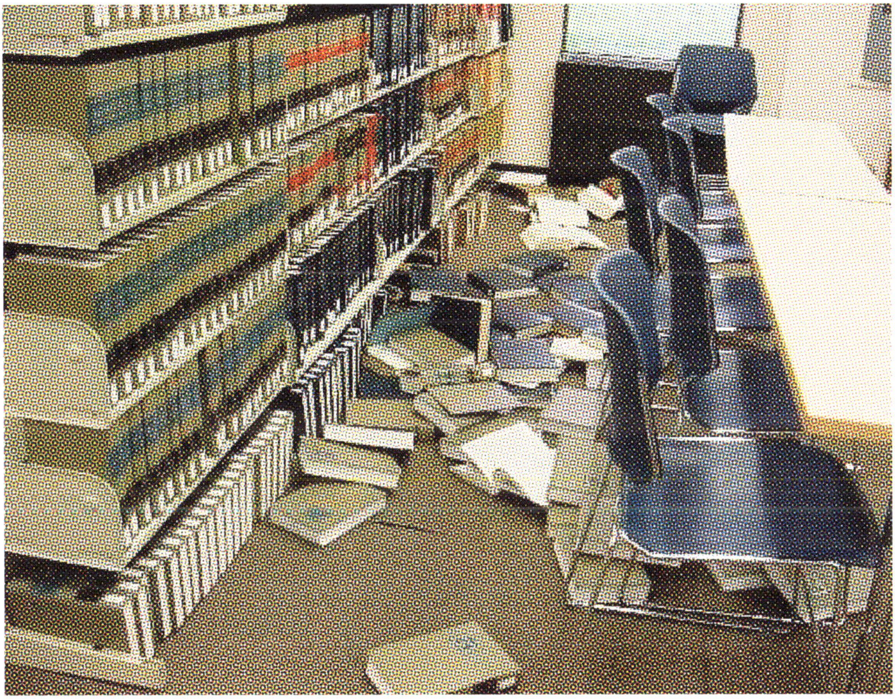

Nisqually, USA

28 February 2001

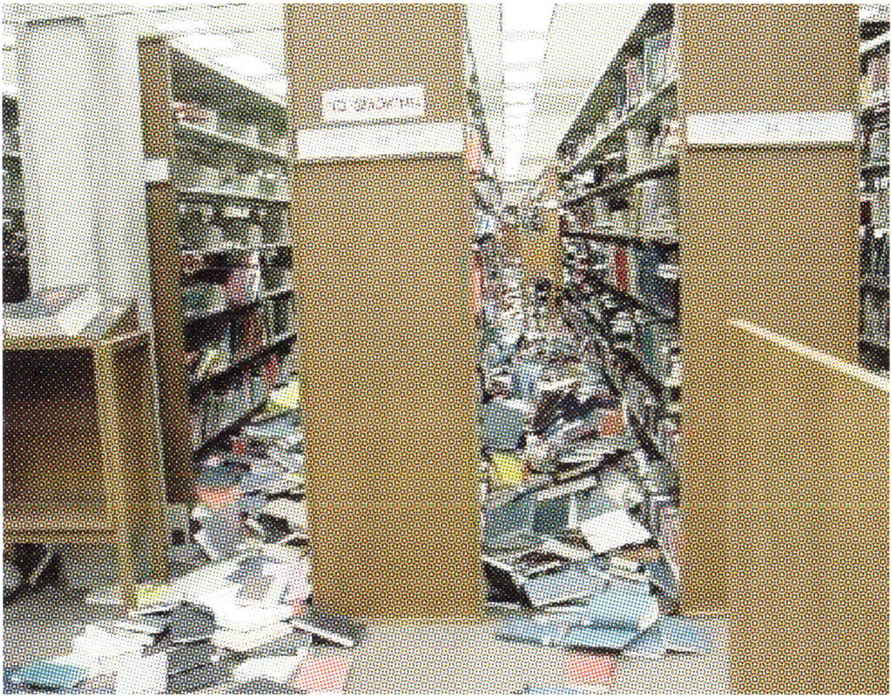

Nisqually, USA

28 February 2001

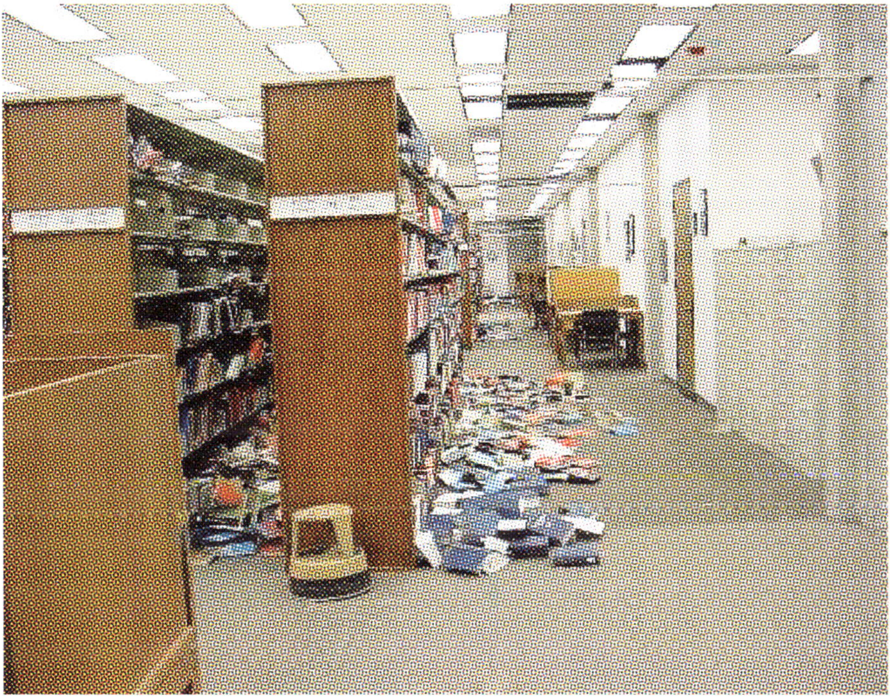

Nisqually, USA

28 February 2001

The Evergreen State College D.J. Evans Library

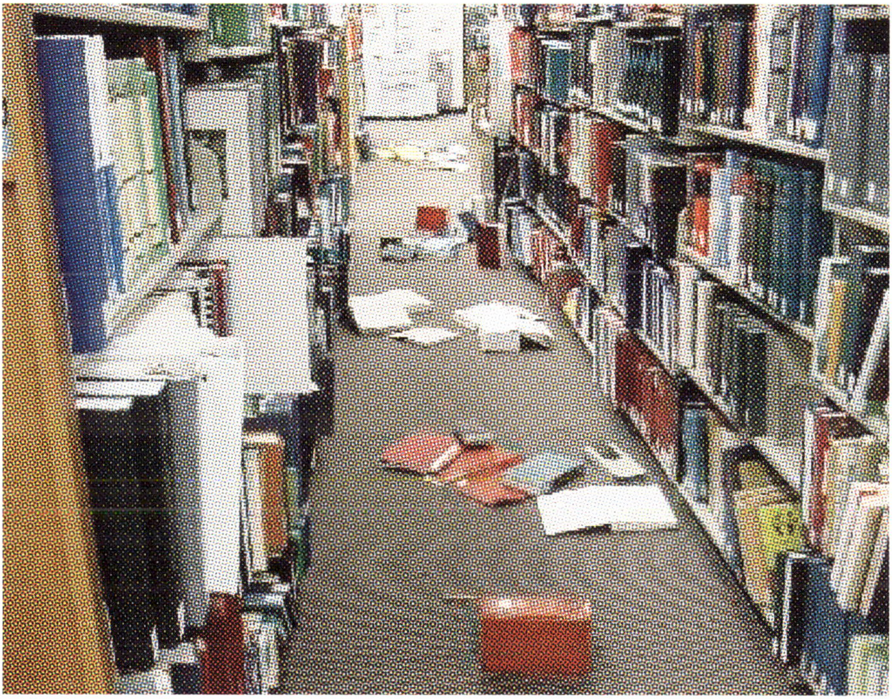

Nisqually, USA

28 February 2001

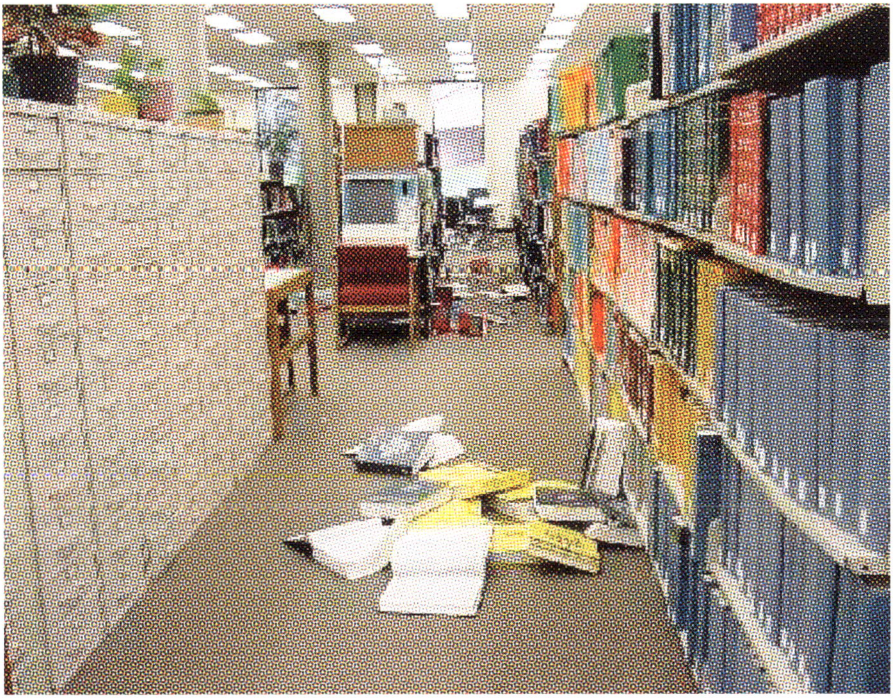

Nisqually, USA

28 February 2001

The Evergreen State College D.J. Evans Library

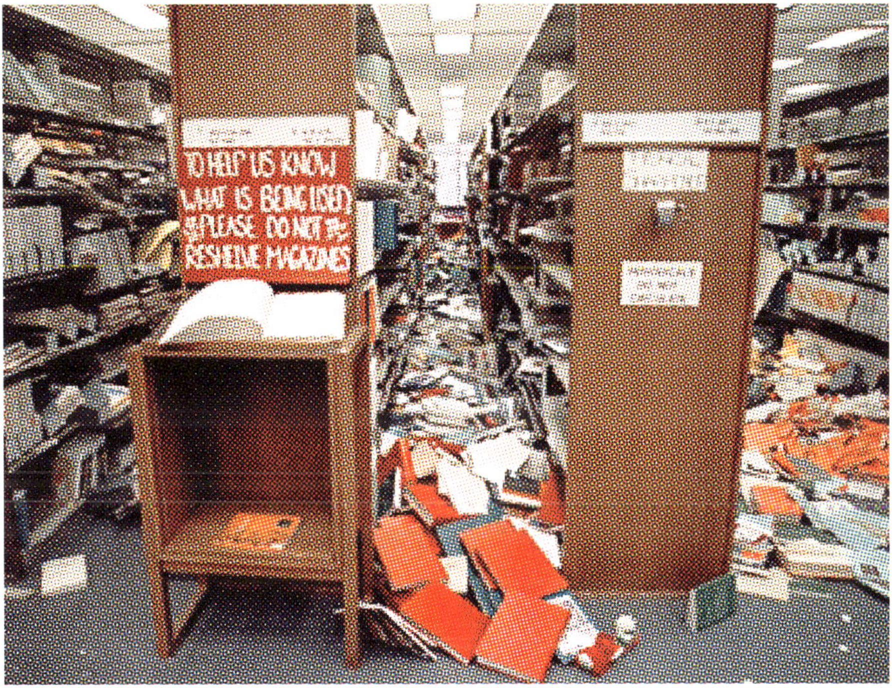

Nisqually, USA

28 February 2001

The Evergreen State College D.J. Evans Library

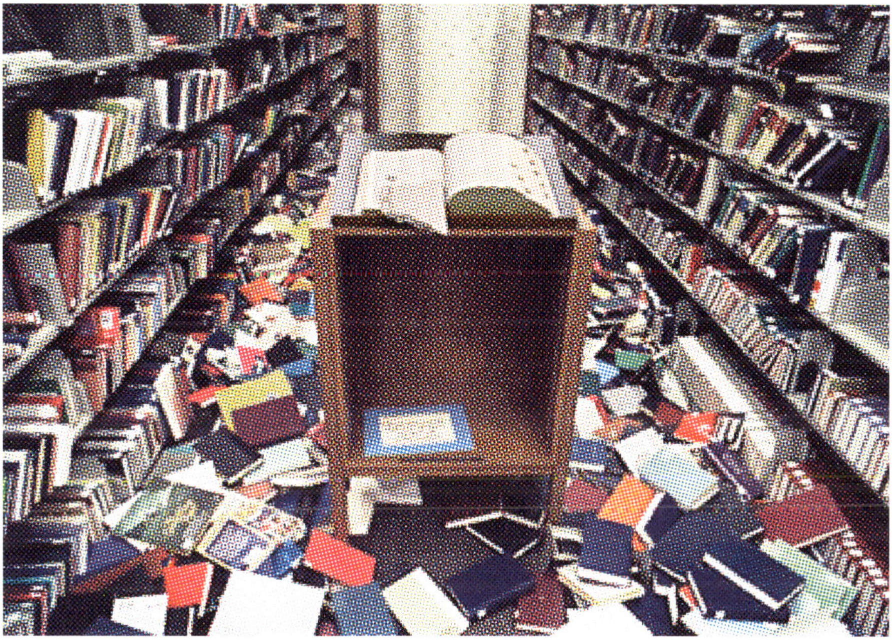

Nisqually, USA

28 February 2001

Dr. Martin Luther King, Jr. Library

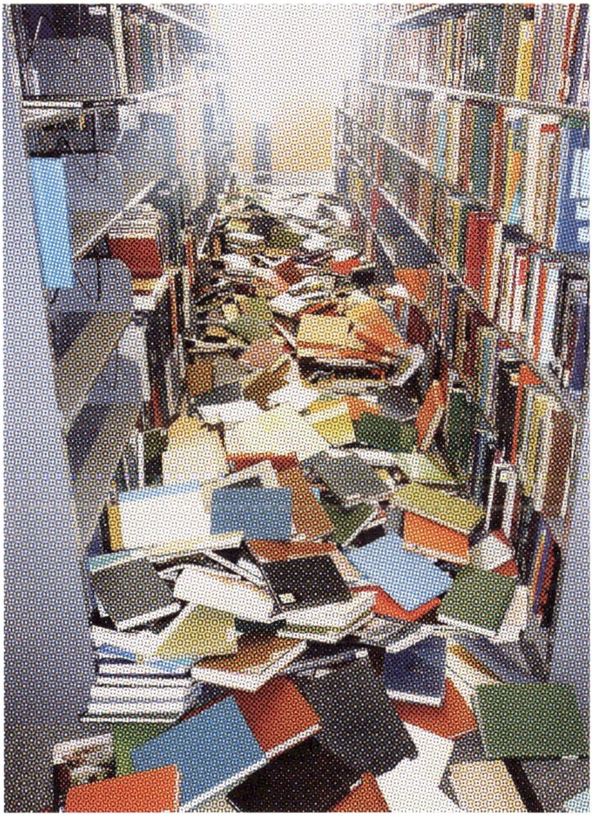

San Jose, USA

30 October 2007

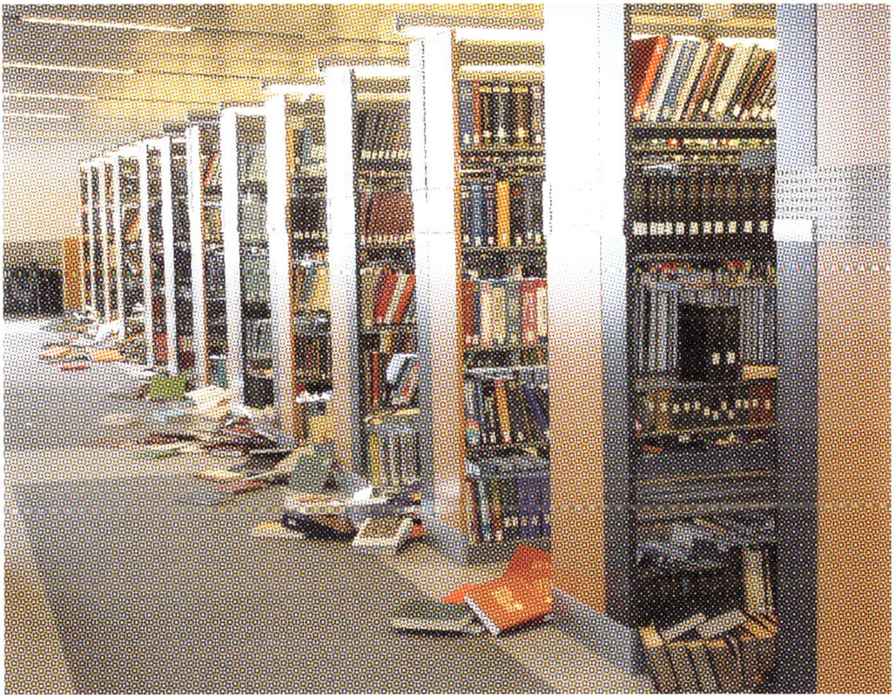

San Jose, USA

30 October 2007

Dr. Martin Luther King, Jr. Library

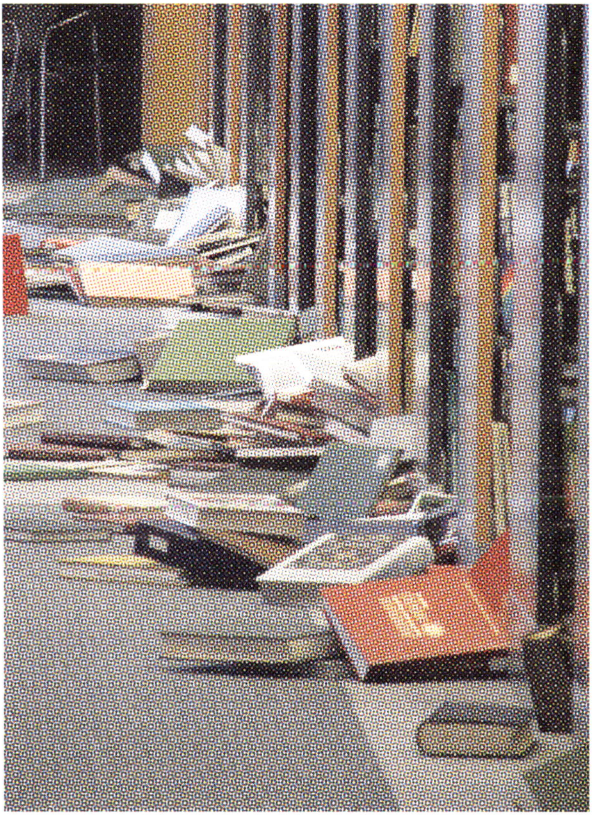

San Jose, USA

30 October 2007

Dr. Martin Luther King, Jr. Library

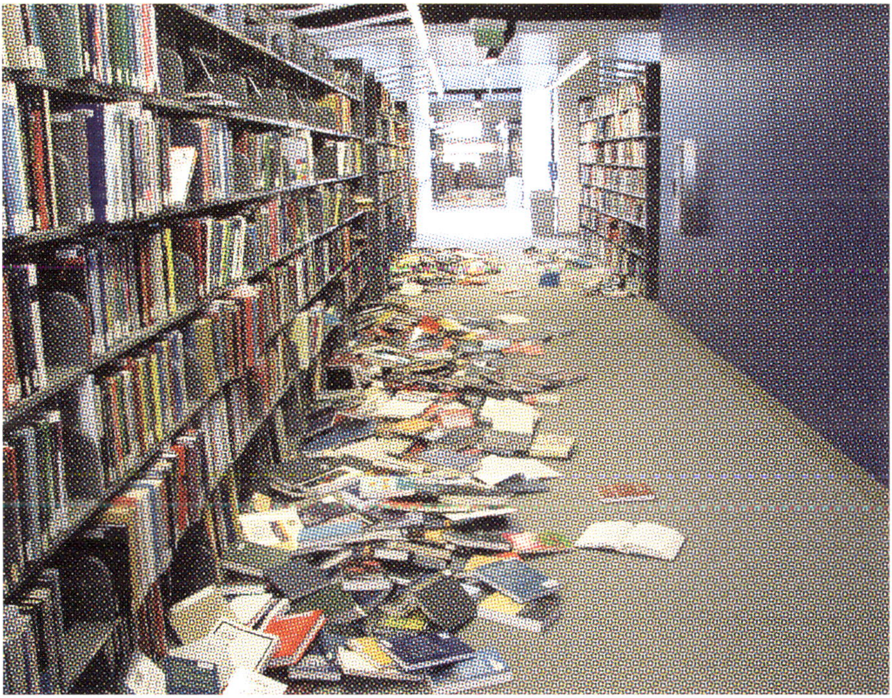

San Jose, USA

30 October 2007

Dr. Martin Luther King, Jr. Library

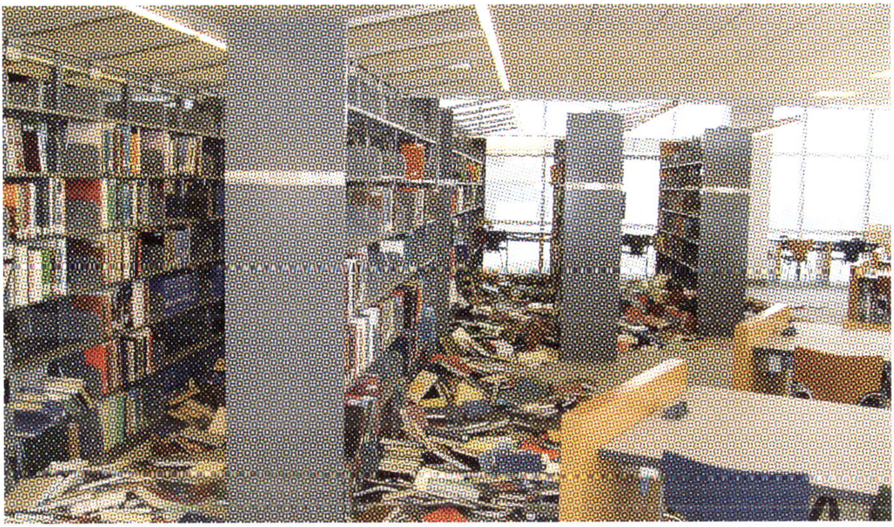

San Jose, USA

30 October 2007

Dr. Martin Luther King, Jr. Library

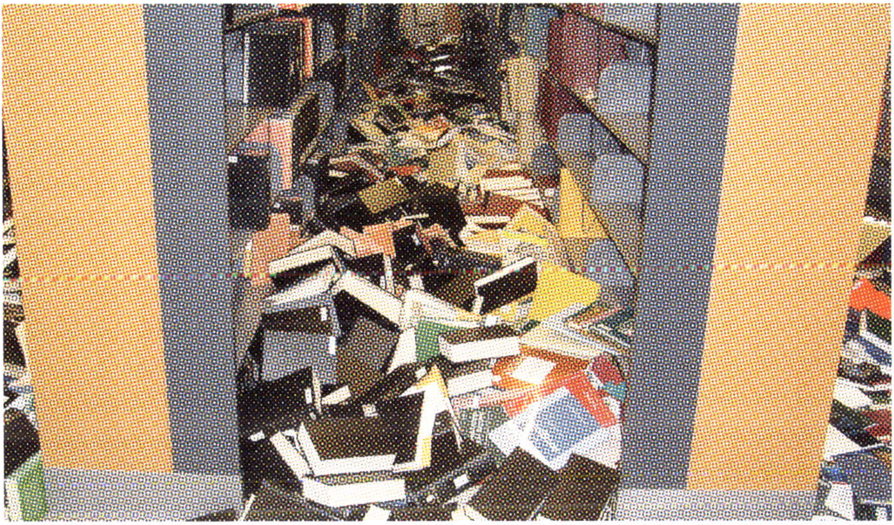

San Jose, USA

30 October 2007

Dr. Martin Luther King, Jr. Library

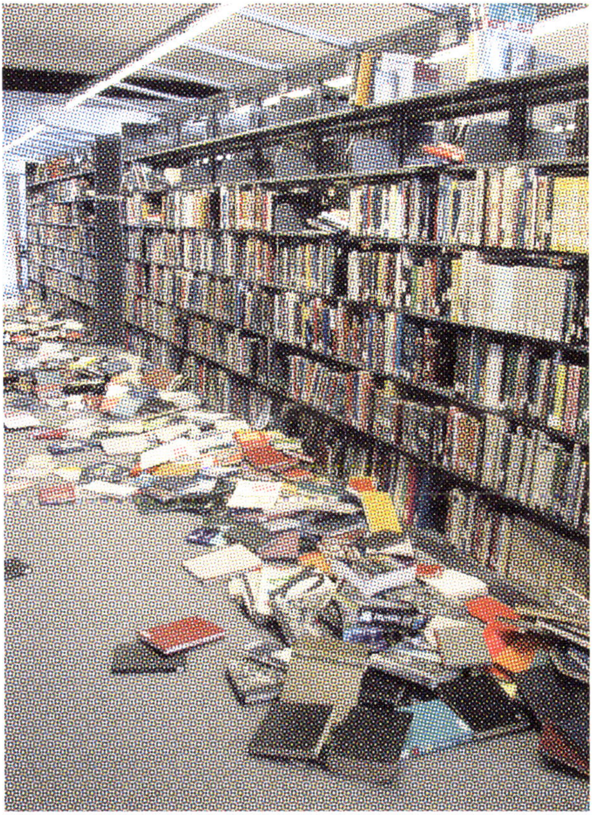

San Jose, USA

30 October 2007

H.B. Williams Memorial Library

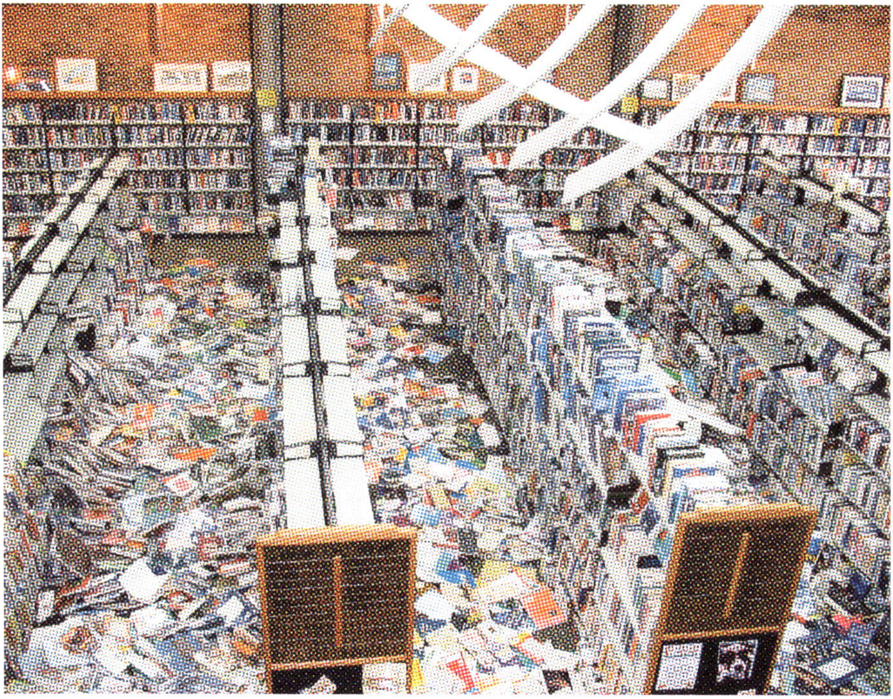

Gisborne, New Zealand

20 December 2007

H.B. Williams Memorial Library

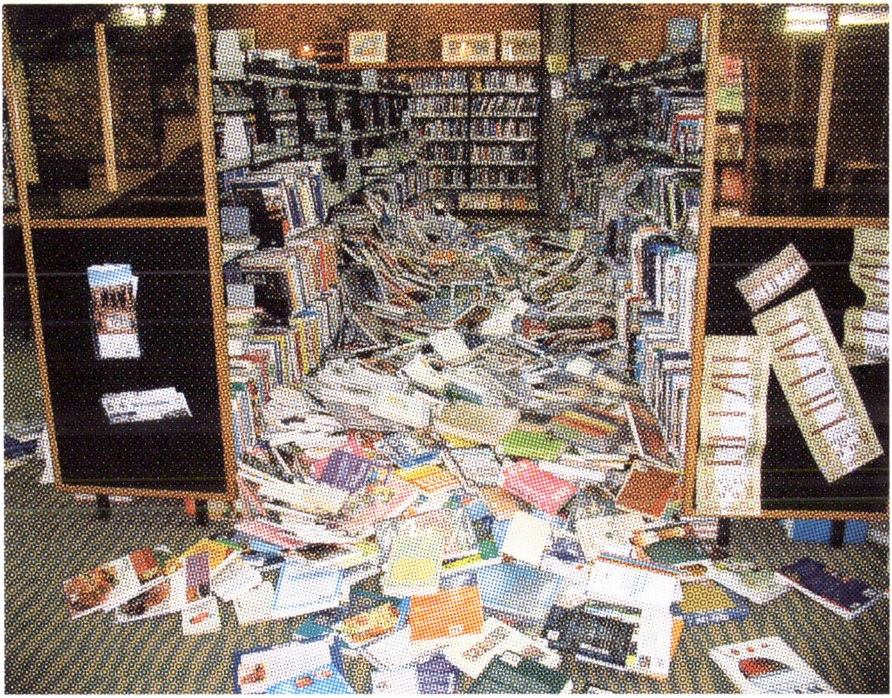

Gisborne, New Zealand

20 December 2007

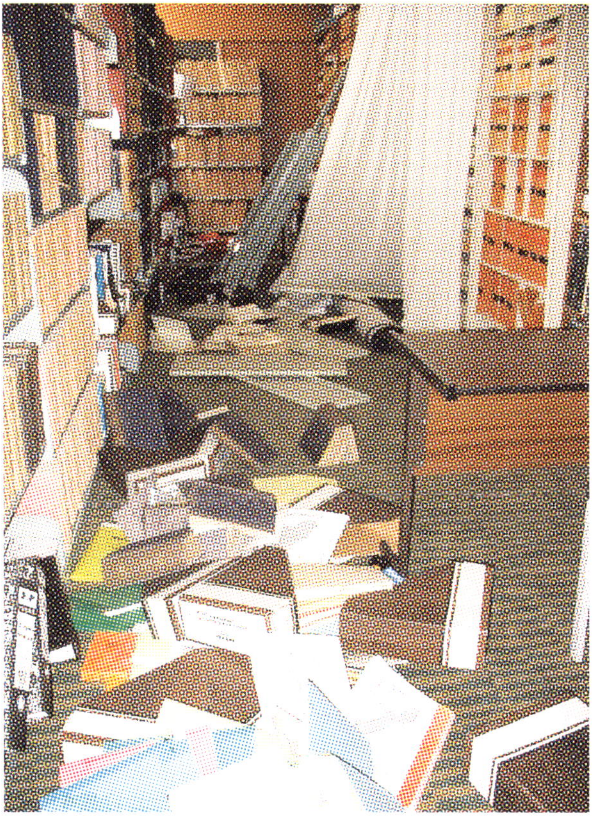

A number of buildings will need demolishing, many homes will need repairs and tonnes of glass and crockery have been lost. Everyone agrees on one thing ... in the words of Civil Defence Controller, Jon Davies 'I think we have actually come out of this one quite luckily,' it could have been a lot worse.

Gisborne, New Zealand

20 December 2007

A Project by
Melissa Dubbin &
Aaron S. Davidson, 2008
Brooklyn, New York

Graphic Design
Francesca Grassi

Published by
onestar press
16 rue Trolley de Prévaux
75013 Paris, France
info@onestarpress.com
www.onestarpress.com

ISBN 2-91535 9-30-X

Edition
/500

Printed at
Cassochrome, Waregem
www.csc.be

Printed in Belgium

Special Thanks
Colleen E. Allen
Dave Balcom
Ami Barak
Christophe Boutin
Kristin Cheney
Carolyn Cole
Susan C. Curzon
Sheryl Davis
Marc Eberhard
Susan Fatemi
Marc Ganzglass
Herrick Heitman
Kathleen Hertel-Baker
Pierre Huyghe
Tabara Katsunori
Bridget Kowalczyk
Betty Mason
Bob Menanteaux
Marie Mestas
Cynni Murphy
Lorraine Oback
Kara Phillips
Mélanie Scarciglia
Randolph Stilson
Kelly Stoker
Arthur G. Sylvester
Pene Walsh

A very special thanks to Francesca Grassi for her enthusiasm and essential contributions to this project.

Images Courtesy
The Seattle Times, Washington
[Page 3]
Photograph by Vic Condiotty

National Information Service for Earthquake Engineering, EERC, University of California, Berkeley
[Page 5]

Anchorage Museum Library and Archives, Alaska
[Page 7 (B69.11.22)]

California State University Northridge, Oviatt Library, Special Collections and Archives
[Page 9, 115, 117]

Los Angeles Public Library, California
[Page 11]

University of California Santa Barbara Libraries
[Page 13]
Photograph by Larry Parsons

National Oceanic and Atmospheric Administration (NOAA) Natural Hazards Image Earthquake Collection at NOAA's National Geophysical Data Center
[Page 15, 17, 21]
[Page 21]
Photograph by J.D. Griggs

Mono Herald and Mammoth Times, Mammoth Lakes, CA
[Page 19]

Inland Empire Libraries Disaster Response Network, University of California, Riverside
[Page 23, 27-49, 55-71]

Stanford News Service, Stanford University, California
[Page 25]
Photograph by Chuck Painter

San Bernardino Valley College, California
[Page 51, 53]
Photographs by Marie Mestas

Santa Monica High School 1994 Yearbook
[Page 73]

Santa Monica Public Library Image Archives, California
[Page 75-113]

Earthquake Digital Disaster Archive of Kobe University, Japan
[Page 119-151]

Seattle University School of Law, Washington
[Page 153-159]
Photographs by Donna Turner

The Olympian, Olympia, Washington
[Page 161, 162]
Photograph by Mike Salsbury, caption Ruth Longoria

University of Washington Fisheries-Oceanography Library, Seattle
[Page 163-167]

University of Washington Physics-Astronomy Library, Seattle
[Page 169-171]
Photographs by Mark McDermott and Peter Shaffer

University of Washington
Engineering Library, Seattle
[Page 173-181]

Evergreen State College Archives
and Special Collections,
Olympia, Washington
[Page 183, 185]
Photographs by Carlos A. Diaz
[Page 187-195]
Photographs by Barbara
Bergquist
[Page 197, 199]
Photographs by Steve Davis

Dr. Martin Luther King, Jr.
Library, San Jose, California
[Page 201-213]
Photographs by Rebecca Feind
and Carole Correa-Morris

H. B. Williams Memorial Library,
Gisborne, Aotearoa, New
Zealand
[Page 215, 221]
Photographs by Dylan Walsh
Munro
[Page 217, 219]
Photographs by Anna
Williamson